Energy Technologies and Economics

Patrick A. Narbel · Jan Petter Hansen
Jan R. Lien

Energy Technologies and Economics

Springer

Patrick A. Narbel
Dept. of Business & Management Science
Norwegian School of Economics
Oslo
Norway

Jan Petter Hansen
Jan R. Lien
Department of Physics and Technology
University of Bergen
Bergen
Norway

ISBN 978-3-319-37946-3 ISBN 978-3-319-08225-7 (eBook)
DOI 10.1007/978-3-319-08225-7
Springer Cham Heidelberg New York Dordrecht London

Cover picture: Mariusz Paździora

Printed on acid-free paper

Springer is part of Springer Science+Business Media (www.springer.com)

Preface

With a global population heading towards 10 billion, ensuring that there is sufficient energy for everyone is likely the greatest challenge humanity faces today. In fact, it is more pressing than other global problems since energy supply, in particular from sustainable resources, is a prerequisite to solving all other major problems. Consider for example the problems of providing food and clean water: Both food and drinking water can be produced in sufficient quantities if enough energy is available for this purpose.

The problem of sustainable global energy supply is an urgent one, for at least three reasons:

i. The growth in population and the increase in the standard of living in developing countries alone call for an increased energy supply.
ii. The depletion of fossil fuel resources (and potentially a crisis when the global demand for oil cannot be met by oil producing countries) will occur in the foreseeable future.
iii. Increased global warming, which is predicted to reduce the planet's ability to host the growing number of inhabitants, is a distinct possibility.

This book is based on several courses on energy-related topics held by the authors at the University of Bergen and at the Norwegian School of Economics (NHH). In particular, the recent master-level course at the NHH, "Alternative Energy Sources in Physical, Environmental and Economical Perspectives," has motivated the authors to create a text that covers the physical principles behind the most common energy sources on the planet, in combination with estimates on how large the various exploitable resources are. These aspects are combined with an original economic analysis on how much the utilization of resources actually costs us today, and how much it could in the future.

This mixture of the physics, technology and economy involved in human energy consumption—the first and only book of its kind—has since evolved into the basis of a popular course in the international MSc program on energy for Economics students. Though primarily intended as a textbook for first-year courses on energy and society, it is also relevant for all interested readers, providing as it does a collection of concrete facts useful in evaluating the often politically biased statements on human energy production and climate. The text is presented at the simplest possible mathematical level, making it also accessible to readers with no background in physics, engineering or economics. At the same time, essential physical concepts are introduced wherever necessary, ensuring that the estimates and predictions are quantitative. For each current and potential future energy source, we examine its physical origin, production technologies, resource considerations, price and environmental pros and cons.

In making this book we have benefited from the knowledge and discussions with numerous colleagues at The University of Bergen and at the Norwegian School of Economics. Also, we are grateful for the professional support from the Springer staff in transforming our initial lecture notes into a complete book.

Contents

1 Basic Physical Processes and Economics 1
1.1 Introduction 1
1.2 Energy 2
1.2.1 Forms of Energy and Important Transformation Processes 5
1.3 Power 11
1.4 The Building Blocks of Matter 14
1.5 Energy and Climate 19
1.5.1 Energy Flux Earth-Sun 19
1.5.2 Radiation Details and the Greenhouse Effect 21
1.5.3 The Carbon Cycle and Enhanced Global Warming 26
1.6 Energy Economics 28
1.6.1 Net Present Value Approach 29
1.6.2 Real Option Approach 31
1.7 Levelized Costs of Energy (LCOE) 34
1.7.1 Capital Costs 35
1.7.2 Operation and Maintenance Costs 36
1.7.3 Fuel Costs 36
1.7.4 Full LCOE Formula 36
1.8 Other Relevant Economical Aspects 40
1.8.1 Introduction to Resource Economics 40
1.8.2 Indirect Costs of Energy 42
1.9 Exercises 47
References 50

2 Fossil Energy Systems 51
2.1 Historical Development 51
2.2 Coal 53
2.2.1 Coal Plants 54
2.2.2 Present Use, Resource Considerations and Forecast 65
2.2.3 Cost 67
2.3 Oil and Gas 71
2.3.1 Basic Properties 71
2.3.2 Hydrocarbon Accumulations 72

2.3.3 Exploration 74
2.3.4 Drilling 76
2.3.5 Production 79
2.4 Enhanced Oil Recovery 84
2.4.1 Oil: Present Use, Resource Considerations and Forecast 86
2.4.2 Some Remarks on the Economic Characteristics of Oil 91
2.4.3 Natural Gas: Present Use, Resource Considerations and Forecast 92
2.4.4 Cost 94
2.5 Carbon Capture and Storage 98
2.6 Peak Oil 103
2.7 Non Conventional Fossil Energy Sources 107
2.7.1 Basic Definition 107
2.7.2 Tar Sands or Extra Heavy Oils 107
2.7.3 Shale Gas 111
2.7.4 Shale Oil 113
2.7.5 Coal Conversion 114
2.7.6 Methane Hydrates 115
2.8 Exercises 117
References 119

3 Nuclear Energy Systems 121
3.1 Historical Development 121
3.2 Basic Nuclear Physics 123
3.3 Nuclear Fission 125
3.4 Nuclear Reactors 127
3.5 Present Use, Resource Considerations and Forecast 130
3.6 Cost 132
3.6.1 Capital Costs 132
3.6.2 Operation and Maintainance Costs 133
3.6.3 Resource Costs 133
3.6.4 Storage Costs 135
3.6.5 Decommissioning Costs 135
3.6.6 Electricity Generation Costs of Nuclear Fission 136
3.7 Nuclear Waste; Safety and Storage 137
3.8 New Nuclear Technology 138
3.8.1 Fast Reactors—Generation IV 139
3.9 Fusion 142
3.10 Exercises 145
References 146

4 Renewable Energy 147
4.1 Historical Development 147
4.2 Solar Power 150
4.2.1 Basic Processes. 150
4.2.2 Area Requirements 155
4.2.3 Present Use, Resource Considerations and Forecast. 155
4.2.4 Cost 157
4.3 Wind Power 159
4.3.1 Basic Processes. 161
4.3.2 Potential and Area Requirements 164
4.3.3 Present Use, Resource Considerations and Forecast. 165
4.3.4 Cost 168
4.4 Hydropower 172
4.4.1 Basic Processes. 173
4.4.2 Present Use, Resource Considerations and Forecast. 174
4.4.3 Cost 176
4.5 Biomass 178
4.5.1 Basic Processes. 178
4.5.2 Potential and Area Requirements 180
4.5.3 Present Use, Resource Considerations and Forecast. 183
4.5.4 Crops for Food or for Fuel? 185
4.5.5 Cost 187
4.6 Other Renewables 189
4.6.1 Geothermal 189
4.6.2 Tides 191
4.6.3 Waves 193
4.6.4 Osmosis 196
4.7 Storage and Energy Carriers 198
4.7.1 High Voltage Direct Current Transmission Lines 198
4.7.2 Batteries 199
4.7.3 Pumping Storage Hydro. 201
4.7.4 Hydrogen. 202
4.7.5 Methanol 204
4.8 Exercises 206
References 209

5 Outlook 211
5.1 Summary of the LCOE of All Technologies 211
5.1.1 Experience Curve 212
5.1.2 Forecasting Future Energy Costs: The Cases of Solar PV and Onshore Wind Power 214
5.1.3 Sensitivity of Solar Power and Onshore Wind Power to Capital Cost Reduction 219

5.2 Energy Policy 220
5.2.1 Introduction 220
5.2.2 Need for Market Intervention 221
5.2.3 Externalities 222
5.2.4 Sensitivity of Selected Technologies to a Cost on Carbon 223
5.2.5 Solving Climate Change via a Carbon Tax 226
5.2.6 Common Policy Instruments Used to Support Renewable Energy 226
5.3 Energy Spending and Efficiency 232
5.3.1 Energy Spending. 232
5.3.2 The Feasibility of Reducing Energy Consumption Through Efficiency Gain 236
5.3.3 Future Energy Alternatives. 238
5.3.4 A Renewable Alternative for Europe? 239
5.4 Estimating the Cost of Future Global Energy Supply 242
5.4.1 Introduction 242
5.4.2 The Value of Developing Scenarios 243
5.4.3 Scenarios 244
5.4.4 Forecasting Future Energy Costs. 246
5.4.5 Results and Discussion 247
5.4.6 Concluding Remarks on Chapter 5 250
5.5 Exercises 250
References 252

Author Biography 255

Chapter 1
Basic Physical Processes and Economics

Abstract This first chapter provides the reader with the background theory on the physics and the economics necessary to fully understand the material presented in the rest of the book. The starting point of the chapter is a discussion of the concept of energy and the definition of important notions related to this concept, such as the various units used in science and engineering. The Earth-Sun energy system, including the famous greenhouse effect, are then presented. Finally, the basic economic terms related to energy plants, pricing and production are discussed.

Keywords Matter · Energy · Power · Climate · Levelized and indirect costs of energy

1.1 Introduction

It can be argued that sufficient energy supply and energy security is the single most important issue for any nation. In a modern society, energy is needed to provide almost any service a citizen usually takes for granted: Electricity for lighting, heating or cooling systems, transportation services and communication, just to mention a few. Therefore, there is an intimate link between the general standard of living in a nation and the energy consumption per capita, provided that the standard of living is reasonably equal for all citizens. For this reason, energy services and energy security are of fundamental importance for the civilization.

In a global perspective, the energy supply should also ideally be *sustainable*. With sustainability, we understand a way of operating which allows future generations to enjoy at least the same standard of living as the current generation. Our global consumption of energy today is not sustainable since it is dominated by fossil fuels (more than 80 %) and current consumption rates imply that existing resources will last for another few tens (oil) or hundreds (coal) of years before they are depleted (BP 2013). At the same time, parts of the world, in particular in Asia, are striving to increase their economic power per capita, which can only be achieved via an increased energy production. The global energy consumption can thus be expected

P. A. Narbel et al., *Energy Technologies and Economics*,
DOI: 10.1007/978-3-319-08225-7_1,

to increase, which will put pressure on the exploitation of fossil fuel resources. A realistic estimate would suggest an increase in energy consumption by the year 2050 by a factor of at least two compared to the present consumption.

In addition, the world population is growing. Predictions for future world population states that it will grow and pass 10 billion people around 2050, and that it may level out between 12 and 15 billion people in the second half of this century (United States Census Bureau 2008). Such an increase can add another factor two to the global energy consumption.

The final issue, apart from the depletion of the resources themselves, is the concern that exhaust emissions from fossil energy sources, mainly CO_2, will cause global warming by an average of 3–5° (Salomon et al. 2007) in a business as usual scenario. Even if these figures are uncertain, it is no doubt that CO_2 contributes to the so-called *greenhouse effect*. The atmospheric greenhouse effect is the ability of the atmosphere to sustain a higher temperature on the surface of the planet than what would be expected with zero atmosphere. An increased average temperature of more than two degrees is believed to lead to a number of negative effects which may result in increased difficulty to live, and even survive, in the near equator regions of Earth. Billions of people can be affected, therefore, reducing CO_2 emissions appears logical. However, reducing human CO_2 emissions in practice and significantly on a global scale has so far not taken place due to the central role played by fossil fuels in the modern society.

The three issues, (1) finite fossil energy supply, (2) a growing population and energy demand and (3) the risk of anthropogenic (i.e.; human based) global warming, taken together contribute to the important question of "**How to supply sufficient sustainable energy to everyone?**" In the following chapters, we will describe the potential, the physics, the economy and the limitations of the most important energy sources of today. But first, the remaining part of this chapter is needed to describe the physical and economical concepts that will facilitate the understanding of the later chapters.

1.2 Energy

The origin of the word *energy* is Greek and refers to "activity" or "being in operation". It was introduced into physics in the 19th century as the mathematical foundation of classical physics, thermodynamics and electromagnetism were developed. Although it is difficult to provide a universal definition of "what energy is", it is sufficient and correct for our purpose to consider energy as a quantity which is conserved within the system in question. Furthermore, energy has the ability to perform work as it transforms from one form to another. Thus, it is correct to state that any process on Earth involves energy conservation. In everyday's life, we continuously experience energy transformation, or the lack of it when we run out of fuel, food or electricity. Note that it is the lack of ability to transform energy which depletes, not the amount of energy in itself, which is constant.

We can divide the various energy forms in two categories: potential and kinetic. Potential energy refers to an energy storage which can be released by certain actions.

The water stored in dams at high altitudes is an example of potential energy. The latter can be released by building a pipeline where the water can flow downwards. As the water flows with an increasing speed, the potential energy is transformed into kinetic energy which can be defined as energy due to a flux or a motion. This kinetic energy is transformed to work when running turbines, and then to electric energy in generators. In gasoline, potential energy is stored in molecules which release energy as they undergo reactions leading to exhaust. The resulting kinetic energy is fast moving molecules which are formed in the chemical reactions. The high kinetic energy of these molecules is utilized as the work needed to push the vehicle forwards.

The important point is that the total amount of energy is the same before and after the process (opening the pipeline or igniting the engine). But after the water has arrived at the power plant, a part of the initial potential energy is transformed into electrical energy and parts of it becomes a random low quality energy transferred to the sea. This low quality energy is sometimes called *anergy* which is an energy form which cannot be further utilized for new energy transformations.

Economists may compare energy to money. Potential energy is cash, it allows you to buy things. During the purchasing process, the money you pay is transformed into another substance, for example a mobile phone. In principle, the mobile phone has the same value as the cash you had before you start using it. With time and use, the value of the phone decreases, but at any time, its real value + the lost value = the initial value.

The higher quality our initial form of energy is, the more work can be performed. For example, a dam creating a 1,000 m fall will transform into more useful energy than a dam with a 100 m fall. We can say that the energy quality of water increases with the height it is stored at.

In building energy transformation systems, or power plants, the quality of the original energy source is extremely important when it comes to the economy of the plant. On the background of energy conservation, it is clear that energy is not *produced* nor destroyed in a transformation process. Still we will use this term since it is common and since, in fact, an increased amount of a certain energy form, like electricity, is generated.

Continuing our analogy with cash, money occurs in many currencies. Any currency can be *transformed* into another currency as long as the exchange rate is known. Correspondingly, energy has a *unit* which is a common "measurement stick" allowing us to be quantitative. Unfortunately there are many sets of units. In this book we will always refer back to the generally accepted system of units called the SI system (Système International d'unités).

The three most basic units in this system are the units for length, mass and time[1]:

- Time—is measured in seconds (only 's' is used in the unit). One second is the duration of 9,192,631,770 periods of the radiation corresponding to the transition between atomic energy levels in an atom of Cesium. This means that the radiation

[1] The SI-systems has four more basic units: the *Kelvin* for temperature, the *candela* for luminous inensity, the *ampere* of electric current and the *mole* for the amount of substance.

Table 1.1 The most common prefixes—and their meaning

Prefix	Symbol	Multiply by	Prefix	Symbol	Multiply by
Zetta	Z	10^{21}	Zepto	z	10^{-21}
Exa	E	10^{18}	Atto	a	10^{-18}
Peta	P	10^{15}	Femto	f	10^{-15}
Tera	T	10^{12}	Pico	p	10^{-12}
Giga	G	10^{9}	Nano	n	10^{-9}
Mega	M	10^{6}	Micro	μ	10^{-6}
Kilo	k	10^{3}	Milli	m	10^{-3}

from this particular atom is similar to a stable oscillating system where the time unit is defined by the number of oscillations.

- Length—is measured in meter (m). One meter is the length of the path travelled by light in a vacuum during a time interval of 1/299,792,458 of a second.
- Mass—is measured in kilograms (kg). One kilogram is defined as the mass of a specific platinum-irridium alloy cylinder kept at the International Bureau of Weights and Measures at Sèvres, France. All other masses are compared to this standard.

The definitions may seem a bit odd. But they are accurate and invariable. For example, the old definition of the meter, which was one ten-millionth of the distance from equator to the North Pole along a specific longitude, is not exact and hard to measure.

All the other quantities have units which are combinations of the basic units. The SI unit of energy is a Joule (J), and has the unit [mass $\times$ length^2/time^2] and one Joule equals [1 kg $\times$ 1 m^2/s^2]. A Joule is a rather small unit at a "human scale". Consequently, the numbers quickly grow very large. On the other hand, at a molecular scale, a Joule is gigantic. Describing energy transitions on this level thus imply very small number of Joules. The large range of values relevant for energy transitions implies that we need to be able to handle the large range of numbers in a meaningful way. Therefore we introduce such words as kilo (k), mega (M), giga (G) to describe a given numbers of product of 10 to multiply or divide the original unit with. For example, "kilo" describes multiplication with 1,000, ie. 1 kJ = 1,000 J. A list of some of the most frequent prefixes and their meaning is given in Table 1.1.

The "food-energy" required for a human each day is often given in the non-SI unit kilocalories (kcal) and known to be around 1,500–2,000 kcal (Willliam et al. 2004). By using the upper value, and using the fact that 1 kcal = 4.185 kJ we see that our daily energy consumption to keep the metabolism stable is around 8 MJ.

Another energy measure useful for describing chemical and nuclear transformation processes, is the electronvolt (eV). One eV is a typical magnitude for energy release or uptake in atoms and molecules and has the very small value $\sim 1.6 \times 10^{-19}$ J. Thus, energy transitions in a single molecule are quite small. A lump of matter of about 1 kg however contains an order of 10^{23} molecules so with such large number of molecules involved in chemical transitions, one realises that the energy released can fast become a few MJ. In the atomic nuclei, energy transitions are of the order of a million times more powerful. Thus if the same lump of matter containing 10^{23}

Table 1.2 Annual energy consumption by region (country) and per capita in 1990 and 2008

Region/Country	1990 (EJ)	1990 per capita (GJ/person)	2008 (EJ)	2008 per capita (GJ/person)
Europe	68.4	144.7	73.4	146.9
USA	80.3	320.4	95.8	313.9
China	36.4	31.7	89.2	67.0
Latin America	14.4	40.3	24.1	51.9
Africa	16.2	25.6	27.7	28.1
India	13.7	15.8	25.9	22.6
The world	368.3	69.8	512.3	76.7

nuclei undergo a nuclear transitions, a TJ[2] of energy can be released almost instantly. This is a vast amount of energy and explains why atomic bombs can cause so much more destruction than a chemical bomb of the same size!

The global annual energy consumption today is about 0.5 ZJ or 5×10^{20} J. In Table 1.2 the distribution of annual energy consumption per year, in absolute numbers as well as in energy consumption per capita is displayed.

We see that the global energy consumption has increased from 1990 to 2008. The average growth for the entire world is about 3 % per year. Secondly, the table shows enormous differences in energy consumption per capita. A US citizen spends about ten times as much energy as an African citizen. When about 8 MJ per day is needed per person in food energy, this amounts to about 3 GJ per person per year. The energy consumption of USA per person is then about 100 times what is needed to eat, while in Africa it is only about 8 times. We also note a relatively stable energy consumption in Europe and in the USA over time, while numbers for China and Latin America show a large increase.

1.2.1 Forms of Energy and Important Transformation Processes

Let us now describe the various forms of energy relevant for the economy.

- Mechanical energy is energy due to masses—either stored as potential energy or in motion as kinetic energy. If a body with mass m has a velocity v, the kinetic energy of the body is:

$$E_{kin} = \frac{1}{2}mv^2. \tag{1.1}$$

  The body may also have potential energy due to its position. If a mass is placed on top of a hill at height h above the ground level, its potential energy will be transferred to kinetic energy when sliding down an ideal frictionless surface.

[2] Remember that 1 TJ = 1,000,000 MJ.

The expression for potential energy on top of the hill is:

$$E_{pot} = mgh. \tag{1.2}$$

The energy conservation implies $E_{kin} = E_{pot}$. The number g is the gravitational acceleration, almost a constant on the Earth's surface and in the atmosphere, and has the value $g = 9.81\,\mathrm{m/s}$.

- Heat, or thermal energy, is energy transfer due to the mechanical energy of atoms and molecules. Since we face very large number of atoms in thermal systems (again $\sim 10^{23}$), we need to express the system in average quantities relevant to the macroscopic world. An important concept is *temperature*. Temperature is a measure of the average kinetic energy of the atoms in the substance. In a gas of non-interacting atoms (ideal gas) the relation between temperature and the atomic or molecular energy is:

$$E_{kin} = \frac{3}{2}kT \tag{1.3}$$

Boltzmann's constant k is here introduced and has the value $k = 1.38 \times 10^{-23}$ J/K. The unit J/K stands for Joules per Kelvin, where a Kelvin (K) is the unit of absolute temperature. From the equations above we see that temperature is always a positive quantity (or zero[3]). All other temperature scales can be derived from the Kelvin scale. The most common in use is the Celsius scale which has its zero at 273.16 K and 100 °C is equal to 373.16 K. Thus a warm summer day of 30 °C corresponds to 303.16 K. We can now precisely define *heat* as an energy flow between systems of different temperatures. Heat is involved almost everywhere in energy systems we rely on; in engines, power plants and as a flux of energy out of our homes on cold winter days.

Note that Eq. (1.3) can be used to estimate the mean velocity of the atoms and molecules constituting our atmosphere. The mass of an oxygen molecule O_2 is equal to roughly 32 times the mass of a single nucleus particle; the proton, $m_p = 1.67 \times 10^{-27}$ kg. From the formulas above we obtain $v = \sqrt{3kT/m} = 483$ m/s at T $= 300\,\mathrm{K}$. This very high velocity should not be mistaken for the *wind velocity*, which is a collective phenomenon to be discussed in the next chapters.

As a result, the total energy of a system of N atoms is proportional to the temperature and the number of atoms:

$$E = \frac{3}{2}NkT \tag{1.4}$$

Heat transfer processes always behave according to the laws of thermodynamics. The first law is nothing but a rephrasing of the energy conservation principle for

[3] Zero temperature is a real theoretical limit. Quantum mechanics forbids any system to reach zero temperature.

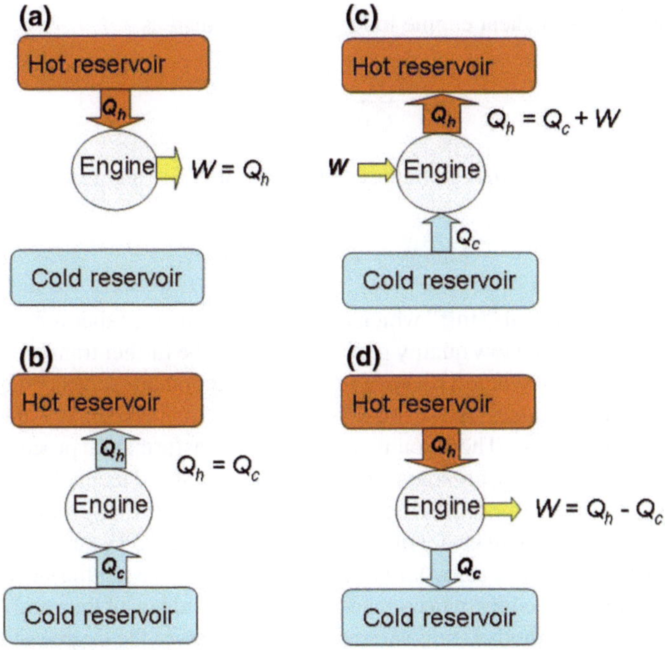

Fig. 1.1 Illustration of forbidden (*left*) and allowed (*right*) heat transfer processes according to the laws of thermodynamics. **a** No engine can transfer heat completely into work. **b** No engine exists which can spontaneously flow heat from a cold reservoir to a hot reservoir. **c** An engine need work to transfer heat from a cold reservoir to a hotter reservoir (refrigirator). **d** A heat engine produces work by taking heat from a hot reservoir (the fuel), but at the same time lose some heat to a cold reservoir (the exaust)

heat transfer processes: change in the internal energy of a closed system is equal to the sum of the amount of heat energy supplied to or removed from the system and the work done on or by the system. The second law states that processes involving heat have a direction: Heat cannot spontaneously flow from a location with low temperature to a location with higher temperature—work is required to achieve this. A related statement is that no process involving heat transfer can take heat from a high temperature region and transfer everything into useful work. Some finite amount of low-quality energy is always simultaneously released into a cold reservoir.

We experience countless examples of processes in agreement with the first and second law of thermodynamics in our everyday life. At home, the refrigerator does not "pump out heat" by itself unless we supply electrical energy. The heat produced in a chemical reaction of a car engine cannot use the entire energy to run the car, since some exhaust carrying part of the heat is always released together with CO_2 and other gases. Allowed and forbidden processes are shown graphically in Fig. 1.1.

The efficiency of the heat engine in Fig. 1.1d is defined as:

$$\eta = \frac{W}{Q_H} = \frac{Q_H - Q_C}{Q_H} \tag{1.5}$$

In fact, the laws of thermodynamics can be generalized to any macroscopic transformation process:

$$Energy = Exergy + Anergy. \tag{1.6}$$

The *exergy* is the useful "stuff" which can be transformed to other forms of energy while the *anergy* is the low quality part which cannot be further transformed, as for example Q_C in the lower left part of Fig. 1.1. The second law states that both terms on the right side of this equation are finite and larger than zero in any macroscopic transformation process. The amount of exergy in a transformation process may also depend on the type of technology used. The more advanced technology, the larger the exergy. As a result more efficient products are obtained. Today a small petrol car engines for example, can easily drive 10 km consuming 0.3 l gasoline. But 20–30 years ago the same sized car would likely have spent 1 l driving the same distance.

A final important concept in connection to the second law of thermodynamics is *entropy*. This a thermodynamic variable which expresses the degree of orderness in a system. Low entropy means high degree of order and often high degree of capability to produce work. A high entropy means that the thermodynamic state of the system has evolved to a state where almost no more exergy can be obtained in any process unless energy is added to the system. An alternative way of expressing the second law of thermodynamics is to say that in any energy transformation process within a macroscopic closed system, the total entropy always increases.

- Chemical energy is the amount of useful energy stored in molecules. When ignited or brought in contact with other molecules, chemical reactions occur which result in heat. An example of a chemical reaction is a reaction between an oxygen atom and a hydrogen molecule, H_2, resulting in a water molecule, $H_2 + O \rightarrow H_2O$. In this process an amount of roughly 5 eV is released. Then, if you take 2 kg of H_2 and mix it with 16 kg O_2, a massive amount of 3 GJ is released.[4] Unless this amount of energy is released in a controlled way, a violent explosion occurs.

  Electrostatic interactions are the origin of chemical energy. Atoms in molecules bond to each other with varying strengths depending on the number of electrons each atom "brings" into the molecule. The energy released in a chemical reaction follows from the different bonding strengths. Well known end products of chemical transformation chains often involve H_2O and CO_2 which have large binding energy (binding strengths). Thus, having produced CO_2 in some reaction energy, you cannot design a new process to transform CO_2 without adding energy.

[4] These quantities are due to the chemical properties of the two combined elements.

Table 1.3 Typical calorific values for some selected fuels in MJ/kg

Product	MJ/kg
Wood	14–17
Fresh fruits	15
Wheat	17
Sugarcane	17–18
Coal	17–34
Ethanol	30
Diesel	45
Gasoline	48
Methane	55

The energy released in chemical reactions is clearly proportional to the mass of the substance undergoing a reaction (since the mass is proportional to the number of reacting molecules):

$$E = C_{th}m \tag{1.7}$$

The proportionality constant, C_{th}, is called the calorific value, and the subscript points towards the fact that the end product of chemical reactions is dominated by heat. Typical calorific values for fossil energy fuels are 20–60 MJ/kg. Typical calorific values for some products are indicated in Table 1.3.

- Electrical energy is the energy carried by the motion of electrons and is transported in conducting materials. The electrical energy can be stored in batteries. However, since a battery capacity is very low compared to the electrical energy required to store it, we can say that electrical energy in practice cannot be stored without transforming it to some other form of energy.

The voltage, or potential difference, across a resitance R is given by Ohm's law:

$$V = RI \tag{1.8}$$

where I is the electrical current represented by the moving electrons, and the electrical resistance R is the effect of collisions between these electrons and atoms in the material. The result is a dissipation of heat.

The transport of electrical energy over large distances usually takes place through high voltage lines. Copper wire is very expensive and in order to save metal and cable weight, the utility companies prefer to use as thin cables as possible. The resistance is inversely proportional to the cross section of the cable, so thiner cables implies higher electric resistance R, and therefore a low current, according to Ohm's law. To compensate for the high resistance, a high voltage must therefore be used when transporting electrical energy over long distances. The consequent low current also minimizes the energy loss due to electrical resistance in the cables because energy dissipated in an electric resistance is proportional to the current squared. In urban areas the voltage is transformed from several hundred kV successively down to a standard 220 V which is the voltage across the power plugs in European houses.

- Nuclear energy refers to the energy released when colliding nuclei transform, or disintegrate by themselves. Again, the origin—as for chemical energy—is a variation of how tight various nuclei bind to each other through the nuclear force. On the scale of the size of the nucleus[5] (10^{-15} m), the nuclear force is much stronger than electromagnetic forces. Thus the energy released in a single nuclear reaction is typically several order of magnitudes larger than the energy released in a single chemical reaction.

  An important formula which explains how energy is released in nuclear reactions is the famous formula by Albert Einstein:

$$E = mc^2 \tag{1.9}$$

  where $c \sim 3 \times 10^8$ m/s is the velocity of light in a vacuum. Measuring the total mass m before and after a nuclear reaction shows that they differ. For reactions releasing energy, the mass after the reaction is smaller than before. The mass change multiplied by c^2 is the amount of energy released during the reaction.

- And last but not least, electromagnetic radiation is the energy carried by massless "particles" called *photons*. These particles are the building blocks of electromagnetic fields and they are the carrier of energy from radiating bodies. The most important radiating body for life on Earth is the Sun, which radiates an almost constant flow of photons (i.e.: Sun rays) radially outwards. Figure 1.2 shows the spectrum of electromagnetic radiation. Note that the visible light is only a small fraction of the entire spectrum. Most of the energy received from the Sun is in the infrared (IR) and ultraviolet (UV) part of the spectrum. The large number of photons hitting Earth per second is the dynamo of almost all derived energy forms on Earth. Oil and natural gas are ultimately solar energy which through photosynthesis in plant material has been transformed to chemical energy and stored for millions of years until humans discovered how to apply it. The energy of the photon is proportional to the frequency, f. The proportionality constant, $h = 6.62607 \times 10^{-34}$ Js, is called Planck's constant, after the famous German physicist Max Planck, who put forward this relation to explain the characteristics of radiation from idealized bodies with a fixed temperature. A photon inside an electromagnetic wave with frequency f then has the energy:

$$E = hf \tag{1.10}$$

  Photons travel in space with the velocity of light, and as they build up an electromagnetic wave, the photon also has a characteristic wavelength given by $c = \lambda f$. Thus we can write:

---

[5] Nuclear forces only act on small distances and rapidly decay outside the size of the nucleus—fortunately!

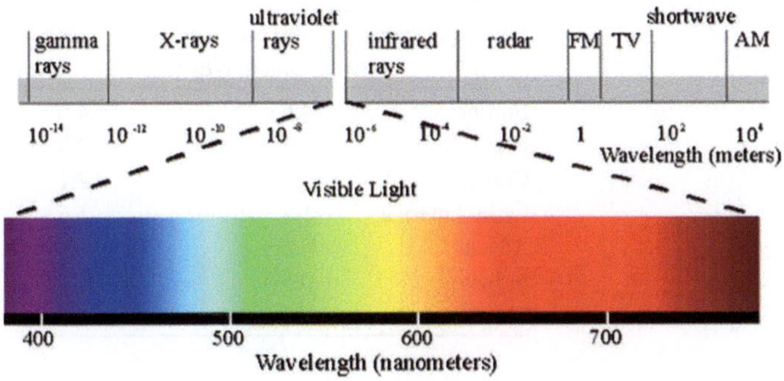

Fig. 1.2 The electromagnetic spectrum—wavelenghts of photons from γ-rays to radiowaves

$$E = hf = \frac{hc}{\lambda} \tag{1.11}$$

Returning to the example of the previous paragraph, we see that photons sent out from nuclei have much smaller wavelengths (and much higher frequencies) than photons radiating from atoms. Short wavelength photons thus carry enough energy to ionize (break up) molecules and atoms and are as such called *ionizing radiation*.

When energy is transformed between any of the types described above, some energy is generally lost according to the laws of thermodynamics. Some transformations are more important than others and an important economical as well as technological question is then how efficient the various transformations are. The transformation of kinetic energy from wind to electrical energy in a wind mill is about 50 % and the transformation process from petrol to mechanical energy in a car is typically around 30 %. In fact the entire world economy can be viewed as a system which is driven by available exergy. A complete alternative theory of economic growth based on this concept has been developed by Ayres and Warr (2010).

1.3 Power

The concept of power is as important as energy itself. It describes how much energy is transformed per unit of time. The average power in a process where an amount of energy E is transformed in a time t is then:

$$P = \frac{E}{t} \tag{1.12}$$

The basic power unit is Watt, where 1 W = 1 J/s. Thus, as a system delivers higher power, more energy is delivered in a given time-frame. Another important power

Table 1.4 Power per capita by region (country) in 1990 and 2008

Region/Country	1990 power per capita (kW/person)	2008 power per capita (kW/person)
Europe	4.6	4.6
USA	10.2	9.9
China	1.0	2.1
Latin America	1.3	1.6
Africa	0.8	0.9
India	0.5	0.7
The world	2.2	2.4

unit is the power density[6] which is the generated power per area. When the area and the power density is known, the total power is found by multiplying power per area by the area.

The power density generated from the various energy forms vary tremendously. From an economic point of view, the energy sources providing the highest power density are often the prefered source of energy. The lower the power density, the larger area and more materials are needed to generate a fixed total amount of power. However, energy sources providing a very high power density, such as nuclear power, may also need very advanced and expensive security systems which may make the picture more complex.

As mentioned previously, the energy intake for humans per day is about 8 MJ. The power our bodies perform at on average, is then:

$$p_{human} = \frac{8\,\mathrm{MJ}}{24\,\mathrm{h}} = 93\,\mathrm{W} \tag{1.13}$$

So an average European human produces energy on average at about 100 W. Most of the energy intake is ultimately released as heat and this is the reason why a group of 10 persons gathered in a small, well isolated room leads to a notable temperature increase. The human body has the ability to perform at much higher power when needed. In le Tour de France for example, the cyclists often perform at 4–500 W for 5 h. At full power, our body can produce 1,000–2,000 W for a short period of some 10's of seconds.

The energy consumption of a nation per year per capita is another interesting and relevant number to look at in the power context. Based on Table 1.2, the average power per person at selected places around the world can be obtained by dividing the annual energy spent per person by one year (=3.1536×10^7 s). This is shown in Table 1.4.

We see that kW per person is a rather good unit to describe a nation's average annual energy consumption. It results in numbers between 1 and 10 which are easy to relate to. And again, as in Table 1.2, the difference between the western and the developing world is striking. In Europe, the power per person is seen to be

[6] More precisely it is called surface power density.

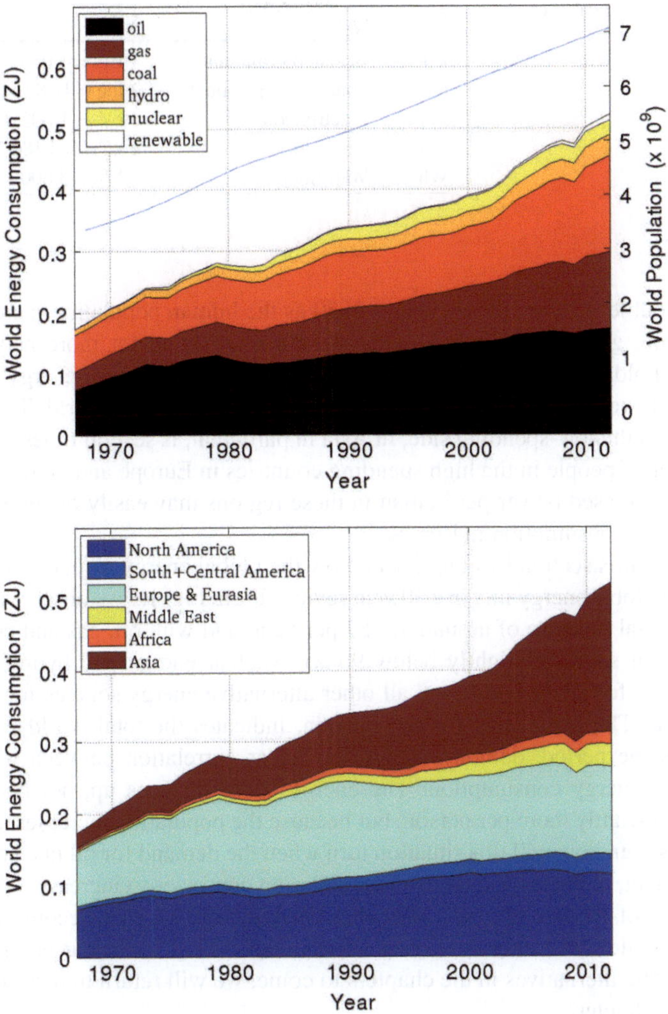

Fig. 1.3 World total primary energy consumption per energy source 1967–2012. (Data from BP.com. Energy consumption from traditional biomass is not included and adds up to about 10 % of the total consumption shown). The *blue line* shows the global population (in units of 10 Gpeople) for comparison (*top panel*). The *lower panel* shows the world total primary energy consumption per region

stable at around 4.6 kW, a relatively large number compared to the few hundred Watts necessary to sustain life for a single person. What is the rest (∼4 kW) of each European's power consumption used for? The answer lies within transport, infrastructure, heating—in short: our civilisation.

Adding up all countries (or continents) we end up with a total power use of about 16 TW. As discussed in the introduction, this number is likely to increase,

Table 1.5 The SI value of various energy units

Unit	Quantity	Value
Btu	British thermal unit	1 Btu = 1 055 J
toe	Tonne oil equivalents	1 toe = 41.87×10^9 J
eV	Electronvolt	1 eV = 1.60218×10^{-19} J
kcal	Kilocalorie	1 kCal = 4 184 J
Wh	Watt-hour	1 Wh = 3 600 J

perhaps double (or even triple) within 2050 as the human population increases and as economic growth allows for higher standards of living for more people. This statement holds even if we manage to increase the efficiencies in energy spending as well as lower the consumption in high-spending parts of the world: The number of people on the low-spending side, in Asia in particular, is several times larger than the number of people in the high spending countries in Europe and North America, therefore increased power per human in these regions may easily offset a decrease of the energy consumption in Europe.

We close this section by considering how the global energy resources contribute to the total global energy mix in a 30 year review. From Fig. 1.3 we observe an almost constant anual increase of around 7.6 EJ per year, and with oil, gas and coal being the dominant sources (slightly below 90 %). Nuclear energy and hydroelectricity both account for about 5–6 % and all other alternative energy sources for only 1 % of the total. The black line, for comparison, indicates the total world population over the same period, demonstrates a very clear correlation between number of people and energy consumption: The energy spending goes up, not because we spend significantly more per person, but because the population increases. A central question is then how will this situation turn when the demand for oil in a near future cannot be met, at the same time as the population continues to increase. Will it lead to a revival of the use of coal? Will alternative energy sources become attractive at a large scale? Will alternative fossil fuels take over? Or nuclear energy? After reviewing the alternatives in the chapters to come, we will return to these questions in the final chapter.

As mentioned briefly, there exist an immense number of energy units: To allow the reader to quickly transform to SI units and related tens and fraction of tens of SI units, we provide here a transformation table for a range of units which are common when discussing various forms of energy tranformation processes (Table 1.5).

1.4 The Building Blocks of Matter

In discussing energy conversion of the various energy sources, it is useful at this point to describe, very simply and in brief, the building of the macroscopic world from its basic building blocks of elementary particles. Our understanding of these

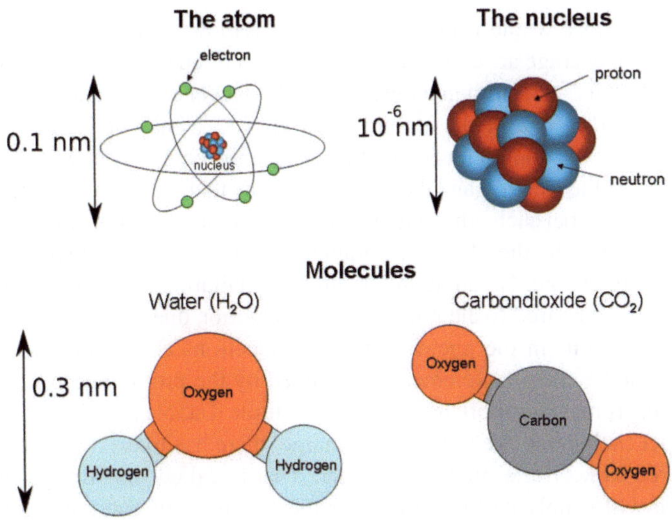

Fig. 1.4 Sketch of a carbon atom (*top*) containing 6 electrons, and its nucleus of 6 protons and 6 neutrons. *Bottom* example of two simple molecules: Water (H_2O) to the *left* and carbon dioxide (CO_2) to the *right*

building blocks went through a rapid development starting late in the 19th century and ended with the development of quantum mechanics by Erwin Schrödinger and Werner Heisenberg in 1926. More than ten years earlier Ernest Rutherford, had established that atoms contained a massive core with a radial extension of only a few femtometers and containing almost all the mass of the atom while the neutral atom itself was about a tenth of a nanometer in size. This atomic core is today known as the *nucleus*. The atomic nucleus is built up of *nucleons*,[7] which are either positively charged *protons* or neutral particles called *neutrons*, see upper part of Fig. 1.4. The neutron and the proton have almost the same mass, $\sim 1.67262 \times 10^{-27}$ kg (proton) and $\sim 1.674929 \times 10^{-27}$ kg (neutron). The negatively charged electron can be viewed as orbiting the nucleus in stable orbits like planets orbiting the Sun. The electron mass is about 1,836 times smaller than the proton mass and the typical radius of the outer orbiting electron in a stable atom is the Bohr radius,[8] $a_0 \sim 0.5 \times 10^{-10}$ m. Hydrogen is known as the simplest atom of all, consisting of only a single electron orbiting its nucleus.

A fundamental property of elementary particles, *charge*, has here been introduced. This is another conserved quantity characterizing all stable elementary particles. The charge of a particle is either positive, zero or negative, and it is quantized in units of $e \sim 1.6 \times 10^{-19}$ C. The unit C here stands for 'Coulomb' and is the unit of charge in the SI system. As the number of electrons, each with a charge of -1.6×10^{-19} C and the number of protons, each with a charge of $+1.6 \times 10^{-19}$ C are the same in all

---

[7] We do not here consider that the nucleons are built of more elementary particles called *quarks*.

[8] After the famous Danish physicist Niels Bohr who put forward the first quantum theory of the hydrogen atom in 1913.

atoms, the atom as a whole is charge neutral. To remove an electron from an atom costs energy, on average around 10 eV. This amount of energy can be provided by a photon which "hits" the atom and destroys itself. The energy the photon originally had is transferred to one of the atomic electrons which may be excited or kicked out of the atom if the photon has more energy than the energy needed to release the binding between the electron and the nucleus. We say that the atom becomes *ionized*.

Equal charges repel each other and opposite charges attract each other. In contrast, all masses attract each other due to the gravitational force, but much weaker than the attraction due to charge. On an atomic scale the gravitational interaction thus plays a minor role as compared to electromagnetic interaction due to charges.

The more protons in the nucleus, the more electrons are bound and stabilized in what is called a chemical element determined by the proton number. In general protons needs as many or more neutrons to constitute a stable atomic nucleus. Since protons repel each other due to the same charge, there has to exist an even stronger binding force to keep the nucleus together. This is called the strong interaction, but here we will simply call it the nuclear force. Since this force is stronger than the electromagnetic force it cannot extent to large distances, otherwise all matter in the universe would have been collected in one large nucleus. So fortunately the nuclear force only works at distances of about a single femtometer and decays very fast to zero on distances beyond that. For this region there are only about an order of 100 chemical stable elements: When about 100 protons are glued together the electromagnetic repulsion reaches the size of the attraction by the nuclear force and as a consequence the nucleus becomes unstable.

Since almost all the mass of an atom is located in the nucleus, the mass of an element is roughly given by the mass of the protons and the neutrons of that element, but not precisely: In the previous section we mentioned that the mass is also related to the binding energy. In general, the more bound an element becomes, the smaller mass per nuclear particle. Therefore, a common new mass unit has been introduced, the atomic mass unit (amu), which is defined as 1/12 of the mass of the carbon isotope which contains 12 protons and 12 neutrons ($^{12}_{6}$C, element number 6, illustrated in Fig. 1.4). The unit 1 amu = 1.66054×10^{-27} kg is almost equal to the proton mass, but slightly smaller. With this definition, the mass of $^{12}_{6}$C is 12 amu and the mass of a single hydrogen atom is 1.00794 amu.

To remove or rearrange nuclear particles in a nucleus in general costs several order of magnitude more energy than to excite or rearrange electrons. This is partly because of the strength of the nuclear force and partly because the nucleons are much heavier than electrons. Only photons with very high energy, γ−rays, with energy a million times the energy of a visible photon, can cause a nuclear transformation. Also, when photons are emitted from the nucleus spontaneously—a process called natural radioactive decay—the released photon energy is in the form of γ−rays. Such photons naturally have enough energy to ionize a large number of atoms almost immediately if they penetrate living material.

Now that we know what atoms are, the next level of complexity is the tendency of atoms to form larger complexes with other atoms, that is formation of single molecules. These molecules again form larger molecules or complexes of molecules in the solid macroscopic substances. The basic mechanism is the same as the mechanism forming atoms: Electromagnetic forces which are attractive on distance. An additional feature with the forces between atoms is that they become repulsive at short distances. In terms of a potential energy diagram as function of distance between two atoms, the potential energy decreases when going from large separation towards shorter separation. At some point, a minimum is reached and at even shorter interatomic distances, the potential energy starts to increase sharply. At this distance of minimum potential energy, typically a few tens of a nanometer (or 2–3 atomic Bohr radius), a stable molecule is formed. A very simple molecule is H_2 consisting of two hydrogen atoms. Examples of two slightly more complex molecules consisting of three atoms are water molecules (H_2O) consisting of an oxygen atom and two hydrogen atoms and carbon dioxide (CO2) consisting of a carbon atom and two oxygen atoms. These two molecules are illustrated in the lower part of Fig. 1.4. These two molecules, in particularly water, are the dominating gases which regulate the surface temperature on Earth due to their special abilities to absorb radiation.

The atoms in a molecule are bound together by *chemical bonds*. There are several types of such bonds, but they all originate from the outermost electrons of the atom. When two atoms combine in such a way that one atom gives one or more of its electrons to the other atom, the bond formed is called a *ionic bond*. A familiar example is sodium chloride (NaCl), which is common table salt. The sodium atom (Na) has a loosely bound outer electron which it easily gives off to chlorine atoms (Cl). The bonding is then caused by the electrostatic (Coulomb) attraction between the two oppositely charged ions. Another very common bond in molecules, is the *covalent bond*. In this case, electrons supplied by either one or both atoms are equally shared, and act as a "glue" holding the two nuclei together. In hydrogen gas H_2, the two atoms are held together by a covalent bond. Slightly more complicated examples are the water molecule (H_2O), sodium dioxide (CO_2) and methane (CH_4). In the latter case one covalent bond is formed between the carbon atom and each hydrogen atom, resulting in a total of four C–H covalent bonds. This is an example of a very common binding between atoms to form a molecule. A third type of bond is the *metallic bond*, where the bonding electrons are delocalized over many atoms. By contrast, in ionic compounds, the locations of the binding electrons and their charges are static.

Solid materials are due to intermolecular bonding formed between two or more (otherwise non-associated) molecules, ions or atoms. Intermolecular forces cause molecules to be attracted or repulsed by each other. Often, these define some of the physical characteristics (such as the melting point) of a substance.

A characteristic feature of matter, from single atoms to the solid state, is that the binding electron can take certain discrete states. The lowest state, in which 99.9 % of the molecules in our bodies and around us exist in at all times, is called the ground state. Then above the ground state there may be excited states with discrete or a band of allowed energies. Between the discrete states (or band of states) nature does not allow the matter to exist. Therefore, atoms, molecules and solid matter can absorb photons with energy corresponding to the difference between discrete levels or discrete bands. The nature of the energy bandwidths then determines how effective the photosynthesis in plants or the electric current production in solar panels can perform.

Atmospheric molecules have the important property that they may simultaneously vibrate and rotate. Both rotational energy and vibrational energy is quantized similarly to electronic energy levels, but the energy separation between two vibrational states are typically 100 times smaller than between electronic energy levels. And there are hundred, sometimes hundred of thousands rotational energy levels per vibrational energy level. For this reason atmospheric molecules can be considered *rotationally hot*, which means that at standard atmospheric temperatures they exist in almost any rotational energy state. Vibrationally the situation is very different. The radiation outwards from Earth contains photon energies such that H_2O and CO_2, which tend to be in the lowest vibrational energy state, is excited to a higher vibrational state. The absorbed energy is then kept for a short time (a few nanoseconds) and then re-emitted in a random direction. Roughly speaking, half of the absorbed radiation is sent upwards, while half of it is sent downwards and then contributing to additional heating of the Earth. This is the origin of the so-called *greenhouse effect*.

In closing, it may be worth pointing out that the greenhouse effect has little resemblance with the actual heating process of a real greenhouse: In the latter the air inside of the greenhouse is heated by the plants inside which absorbs the incoming solar radiation. The transfer of energy to the air molecules takes place mainly by collisions between air molecules and plants or the window. By confining the air inside the greenhouse it is not allowed to escape and the temperature is kept higher than the outside. This process of energy exchange, in which the air moves around, is called *convection*. By opening the window in a real greenhouse air will exchange with the surroundings and the temperature will go down. In the atmospheric greenhouse effect the air molecules are heated by the radiation from Earth. The outward radiation in a real greenhouse has almost no influence on the temperature inside it.

This very brief description of the world around us is necessary to understand what goes on in several basic energy conversion technologies: Nuclear energy plants extract energy from nuclear reactions which result in new and more stable nuclei. Fossil fuels extract energy from chemical reactions which through burning converts pure carbon (coal) or hydrocarbons (oil and gas) and diatomic oxygen molecules to CO_2 and other waste molecules. One of the main reasons for a positive energy release in the process is the fact that CO_2 is a very abundant molecule. Once it is formed there exists no known process which simply extract energy from this molecule in return of even more abundant molecules. In biofuels, energy is generated through the photosynthesis. Here photons from the radiation of the Sun are absorbed and used for breaking up the CO_2 molecule. Hydrocarbons with potential for releasing energy again are formed while oxygen molecules are released back to the atmosphere. More detailed processes related to specific energy transfer processes will be discussed in the forthcoming chapters.

Several constants of nature were introduced in this section. For an overview and for fast lookup we repeat them all in Table 1.6.

Table 1.6 Constants given with five decimal digits (approximate value)

Constant	Quantity	Value
e	Electron unit charge	$e = -1.60218 \times 10^{-19}$ C
m_e	Electron mass	$m_e = 9.10938 \times 10^{-31}$ kg
m_p	Proton mass	$m_p = 1.67262 \times 10^{-27}$ kg
amu	Atomic mass unit	amu = 1.66054×10^{-27} kg
a_0	Bohr radius (atom size)	$a_0 = 0.529177 \times 10^{-10}$ m
h	Planck's constant	$h = 6.62607 \times 10^{-34}$ Js
c	Speed of light (in vacuum)	$c = 2.99792 \times 10^{8}$ m/s
k	Boltzmann's constant	$k = 1.38065 \times 10^{-23}$ J/K

1.5 Energy and Climate

When assessing renewable energy forms, it is important to know the main facts regarding the driving mechanisms of the climate balance on Earth. This section aims at providing an overview of the basic processes behind the energy balance of our planet. An extra bonus will be a general understanding of the atmospheric greenhouse effect and the mechanisms behind global warming.

1.5.1 Energy Flux Earth-Sun

Our solar system has a very fortunate position in the outer part of our galaxy. And within our planetary system, the Earth has an ideal size and is located at an optimal distance from the Sun regarding temperature for development of water and carbon-based life forms. Furthermore, our single moon stabilizes the rotational axis of the Earth, securing relatively stable conditions with respect to latitude. On top of this, the large outer planets, in particular Jupiter, act as a gravitational hoover on objects which would otherwise penetrate into the inner part of the solar system and increase the frequency of devastating collisions with Earth. All these aspects, and others, have secured sufficient time and stable conditions for nature to develop advanced life forms on our planet. The driving dynamo behind this development is the Sun, which together with its orbiting planets, are now about 4.5 billion years old, halfway in the lifetime of an intermediate sized star. As seen in Table 1.7, the Sun is a massive energy source: the equivalent of annual human global energy production is released each microsecond!

The energy results from nuclear processes giving rise to an outgoing average radiation flux of 4×10^{30} W. The fraction of the disc area of the Earth divided by the total area of a sphere with radius equal to the Earth-Sun distance, R_{ES}, is the amount of power which hits the top of our atmosphere (see Fig. 1.5).

$$P_{in} = P_{Sun} \frac{\pi R_E^2}{4\pi R_{ES}^2} = 1.80 \times 10^{17}\ W = 180{,}000\,\text{TW} \tag{1.14}$$

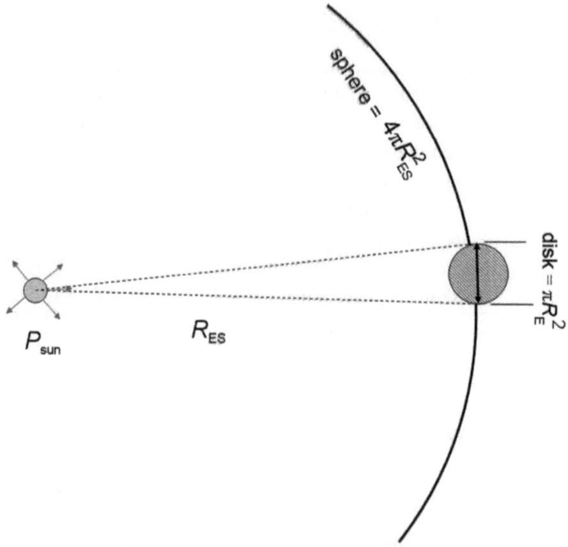

Fig. 1.5 Illustration of the fraction of power from the Sun which hits the Earth

Table 1.7 Some characteristic values for the Earth and the Sun. The distance between the Earth and the Sun is about $R_{ES} = 1.50 \times 10^{11}$ m

Surface temperature	Sun 5,800 K	Earth 288 K
Mass	2×10^{30} kg	5.97×10^{24} kg
Radius	7×10^{8} m	6.37×10^{6} m
Power	4×10^{26} W	1.74×10^{17} W

Note that the energy flux is in orders of magnitude 10^4 times larger than the total global power needed for human life. Therefore, the annual global human energy production is delivered to the Earth from the Sun in less than an hour! About 30 % of the incoming flux is however directly reflected, from clouds (20 %), from the rest of the atmosphere (6 %) and from the ground (4 %). The remaining part of the incomming energy is absorbed by the ground (about 50 %) or by the atmosphere (about 20 %). The absorbed energy is again radiated from the Earth's surface and atmosphere, but then in the infra red region of the electromagnetic spectrum, such that the incoming energy flux is balanced by the outgoing flux. Otherwise the Earth would be warmed up continously.

On average, the power from the Sun onto a square meter of the Earth can be estimated to about 240 W/m^2. This number varies with latitude: At equator it is about twice as large, and much lower at the poles.

The heating of the Earth by the incoming radiation results in outward radiation given by the average surface temperature T_g of the ground. The energy production in the core of the Earth and nuclear processes in the crust also contribute to the temperature, a total of about 32 TW. This geothermal reservoir is indeed enormous, and is estimated to about 10^{28} J. This means that the geothermal reservoir is sufficient

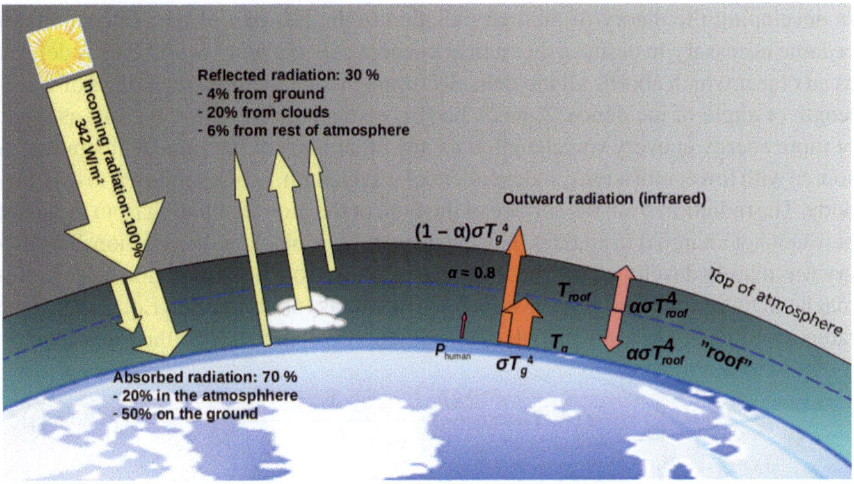

Fig. 1.6 Sketch of energy fluxes in and out of the Earth-Atmosphere system

to cover our present energy production for about 20 million years if it could be extracted. In addition, the gravitational interaction between the Moon (and Sun) and Earth causes tides. As a result, the Earth rotates gradually slower, and the Moon-Earth radius increase with a few cm per year. About 3 TW is finally converted into outgoing radiation due to this phenomenon. A factor which also affect the Earth temperature is the human energy production, which ultimately also ends up as heat. It totals about 16 TW where most of it comes from the processing of fossil fuel. The total power contributing to Earth ground temperature will at equilibrium equal the emitted power from the Earth surface, otherwise the temperature would be constantly increasing.

1.5.2 Radiation Details and the Greenhouse Effect

An object at any temperature emits electromagnetic radiation called *thermal radiation*. The characteristics of this radiation depends on the temperature and properties of the object. At low temperature, the wavelengths are mainly in the infrared region (see Fig. 1.2), and hence the radiation is not observed by the human eye. As the temperature of the object increases, the object eventually begins to glow red, in other words enough visible radiation is emitted so that the object appears to glow. At sufficiently high temperatures the object appears to be glowing white. Careful study shows that, as the temperature of an object increases, the thermal radiation it emits consists of a continous distribution of wavelenghts from infrared, visible, and ultraviolet portions of the spectrum shown in Fig. 1.2.

On the other hand, radiation falling on any object will be partly absorbed, again depending on the wavelengths of the radiation and the temperature of the object.

In developing the theory of thermal radiation in the last part of the 20th century, it became necessary to define a theoretical concept called a *black body*. This is defined as an object which aborbs all incident electromagnetic radiation, regardless of wavelength or angle of incidence. A black body is also an ideal emitter: it emits as much or more energy at every wavelength than any other body at the same temperature. A source with lower emissivity, independent of wavelength, is often referred to as a gray body. The radiation from the surface of the Sun, or the glow emanating from the space between hot charcoal briquettes, has the properties of black body radiation. The theory was mainly developed by Max Planck and is based on the assumption that electromagnetic radiations can only exist in energy quantas (photons) (see Eq. (1.10)). The radiation intensity per wavelength of the radiation will then follow the *Planck curve*:

$$I(\lambda, T) = \frac{2\pi hc^2}{\lambda^5} \frac{1}{\exp\left(\frac{hc}{\lambda kT} - 1\right)} \tag{1.15}$$

Here λ is the radiation wavelength, T is the temperature of the body and the constants h, c, k were defined in Sect. 1.2 and listed in Table 1.6 of Sect. 1.4.

For any temperature of the body, the Planck curve starts up as monotonically increasing with increasing wavelength, then reaches a maximum where the wavelength at maximum radiation intensity follows the *Wien's displacement law*, $\lambda_{max}T = 2.298 \times 10^{-3} mK$. At higher wavelengths, $\lambda > \lambda_{max}$ the intensity rapidly falls towards zero. From this, we see that the Sun and the Earth both emit radiations, but because of the large temperature difference, the radiation from the Sun has a much shorter characteristic wavelength than the radiation from the Earth. In agreement with the second law of thermodynamics, the Earth transforms short wavelength- to long wavelength radiations. A large fraction of the radiation from the Sun is visible (that is why we can see it!) while the radiation wavelength from Earth is typically in the infrared part of the spectrum, which our eyes cannot detect. Figure 1.7 shows the Planck curve for incoming radiations from the Sun at $T = 5{,}800$ K and outgoing from the Earth at $T = 288$ K.

The total radiation from the body at temperature T above is found by integrating over all wavelengths:

$$\begin{aligned}\frac{P}{A} &= \int_0^\infty I(\lambda, T)d\lambda = 2\pi hc^2 \int_0^\infty \lambda^{-5} \frac{1}{\exp\frac{hc}{\lambda kT} - 1} d\lambda \\ &= \left(\frac{2\pi k^4}{h^3c^2} \int_0^\infty \frac{x^3}{e^x - 1} dx\right) T^4\end{aligned} \tag{1.16}$$

The last integral appears by substitution of $x = hc/\lambda kT$ and can be shown to be equal to $\pi^4/15$. We now collect all constants in a single number, $\sigma = \frac{2\pi^5 k^4}{15h^3c^2} = 5.67 \times 10^{-8}\text{W/m}^{2\text{K}^4}$ and express the *Stefan-Boltzmanns law* for the total power emitted:

$$P = \sigma A T^4 \tag{1.17}$$

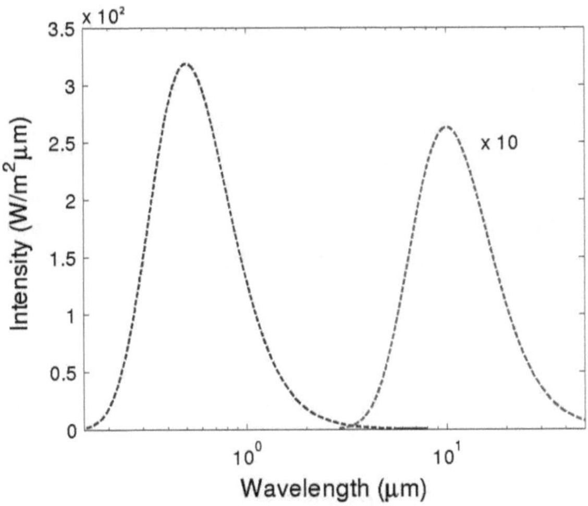

Fig. 1.7 Energy flux per wavelength (*Planck curve*) per square meter from the Sun at 5,800 K towards the Earth surface (*blue curve*) and from the Earth surface at 288 K (*red curve*). The latter is magnified a factor 10 times

By comparing with the known power of the Sun and its area we can compute the surface temperature of the Sun to be about 5,800 K.

Neglecting for a moment the absorption of outward radiation in the atmosphere, energy conservation at the Earth surface at any time imply that the incoming power radiated from the Sun + the relatively smaller contribution of the outward power radiated from the Earth centre, the power induced by the tides and the power generated by humans all add up to the outward radiation power from the Earth surface, characterized by an (average!) ground temperature T_{ground}:

$$P_{total} = P_{in}(1 - S) + P_{Earth} + P_{tide} + P_{humans} = 4\pi R_E^2 \sigma T_{ground}^4 \qquad (1.18)$$

Here S is the fraction of the incoming radiation from the Sun which is reflected. The reader should take note of how small the three last terms amount to compared to the first term on the left side of the equation. Solving this equation with respect to the ground temperature, we obtain an average ground temperature of $T_{ground} = 256$ K or a freezing $-17\,^{\circ}$C. This is more than 30 K lower than the true average global temperature $T_{ground} = 288$ K. The error of the equation above is to neglect the fact that the atmosphere absorbs a large fraction of the outgoing radiation from Earth. We will therefore treat the atmosphere schematically in a model where the upper atmosphere (the troposphere) is replaced by a "roof" which is transparent for the radiation from the sun but absorbs part of the outward radiation from the ground (the Earth's surface). The roof is then a layer which radiates both outwards and inwards, half of the power in either direction, and it must be in thermal equilibrium with a constant temperature T_{roof}. This is expressed by:

$$\alpha 4\pi R_E^2 \sigma T_{ground}^4 = 2\alpha 4\pi R_E^2 \sigma T_{roof}^4 \tag{1.19}$$

Here we have assumed that the "roof radius" is equal to the Earth radius which is a good approximation. The left side of this equation represents the energy supplied to the roof as absorbed thermal radiation from the ground and the right side is the radiated power from the roof. The factor 2 is due to the fact that this radiation is both outwards and inwards. α is the *absorptance* which is the ratio of the radiation absorbed by a surface to that incident upon it. For a black body $\alpha = 1$.

From Eq. (1.19) we find:

$$T_{ground}^4 = 2T_{roof}^4 \tag{1.20}$$

The flux into the Earth in Eq. (1.18) now has to be modified by the additional downward power emitted by the roof:

$$P_{total} + \alpha 4\pi R_E^2 \sigma T_{roof}^4 = 4\pi R_E^2 \sigma T_{ground}^4 \tag{1.21}$$

Eliminating T_{roof} from Eqs. (1.20) and (1.21), we find an expression for the ground temperature:

$$T_{ground} = \left(\frac{P_{total}}{2\pi(2-\alpha)\sigma R_E^2} \right)^{1/4} \tag{1.22}$$

Setting $\alpha = 0$ gives again $T_{ground} = 256$ K since it corresponds to zero absorption of energy in the atmosphere. Setting $\alpha = 1$ corresponds to complete absorption. This leads to $T_{ground} = 301$ K, about 13 K more than the global average. The present average temperature $T_{ground} = 288$ K is obtained with $\alpha = 0.81$, i.e. about 80 % of the outgoing radiation from the ground is absorbed by the atmosphere. The value of α can be changed by changing composition of the atmosphere. In fact, all properties of the atmosphere is included in this parameter, so one can imagine that this model is the simplest possible to describe the greenhouse effect. Note also that the model leads to a lower temperature of the atmosphere compared to the T_{ground}. From relation 1.20, we find $T_{roof} = 0.84 \cdot T_{ground} = 242$ K, or -31 °C. Compare this number to the outside temperature announced by the captain next time you fly!

Let us now take a look at how atmospheric H_2O and CO_2 absorb ingoing and outgoing radiations (see Fig. 1.8). Again the figure shows the Planck curve for incoming radiation but now, it includes the amount of energy reaching the ground from the sun as a thick blue line and the radiation which is transmitted directly into space from Earth as a thick read line. In total, about 30 % of the incoming radiation is directly reflected, partly as it reaches the ground, and partly as it collides with atmospheric molecules especially in the clouds (see Fig. 1.6). Finally, about 30 % of the incoming raditation is emitted directly to space from Earth. The incoming radiation is partly absorbed by ozone (O_3). The other process, called Rayleigh Scattering, is the main reflection mechanism of incoming radiation from the clouds.

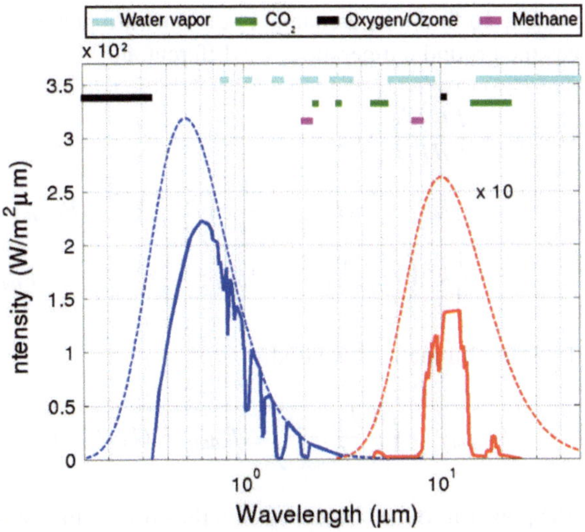

Fig. 1.8 Planck curves of the Sun (5,800 K) and the Earth (288 K). The *broken blue line* shows the spectrum from the sun at the top of the atmosphere and the *thick blue line* the part of the spectrum, about 70 %, penetrating to the ground. The *red broken line* is the spectrum of the *Planck curve* from ground and the *thick red line* is the 19 % fraction escaping directly to space according to Eqs. 1.20–1.23. The absorption wavelengths of some major climate gases are indicated as *thick parallel lines* in the upper part of the figure

Considering the outgoing radiation the main climate gases are CO_2 and H_2O. Figure 1.8 illustrates that almost all the lower part of the Earth's Planck spectrum and the higher part is absorbed by H_2O. In the higher part, around 15–20 μm where the water absorption is less strong, CO_2 is a very effective radiative absorber. Higher fractions of CO_2 in the atmosphere will allow for more absorption of outgoing radiation at wavelengths where the absorption is not already 100 %. It is quite easy to show that a doubling of the present concentration of CO_2 will lead to a global average increased temperature amounting to 1.3 K Thomas and Stamnes (2002). Then, more advanced *climate models* which include couplings between the various concentration of climate gases, the land and the ocean suggest a much more dramatic effect, acknowledged by the IPCC (Salomon et al. 2007). Of other climate gases, we can mention that methane (CH_4) has a strong absorption in the region where Earth radiates, and it is far from saturated. Increasing the methane concentration in the atmosphere by a fixed amount will have a correspondingly much stronger direct effect than increasing the carbon dioxide concentration by the same relative amount.

We can study the potential contribution to climate change from any source within the following model by differentiating Eq. (1.22). First, write:

$$T_{ground} = kP_{total}^{1/4} \tag{1.23}$$

where k is a constant. To find how a change in the outward power from the Earth's surface influences the ground temperature, we differentiate:

$$\Delta T_{ground} = \left(\frac{dT}{dP}\right)\Delta P_{total} \tag{1.24}$$

$$= \frac{1}{4}kP^{-3/4}\Delta P_{total} = \frac{1}{4}k\left(P_{total}^{1/4}\right)^{-3}\Delta P_{total} \tag{1.25}$$

$$= \frac{1}{4}k^4\left(kP_{total}^{1/4}\right)^{-3}\Delta P_{total} = \frac{1}{4}k^4 T_{ground}^{-3}\Delta P_{total} \tag{1.26}$$

Applying Eq. (1.23) to eliminate k^4 we find:

$$\Delta T_{ground} = \left(\frac{1}{4}\frac{T_{ground}}{P_{total}}\right)\Delta P_{total} = G\Delta P_{total} \tag{1.27}$$

A change in the power into the climate system thus results in a proportional temperature change. The term ΔP_{total} is called the *radiative forcing* due to some emitting power source on the ground while G is a constant. If we insert the contribution to the temperature due to human, power consumption, we obtain the very small number $\Delta T_{ground} \sim 5 \times 10^{-3}$ K, which is almost nothing.

However, since our energy consumption also produces a change in climate gas concentration, it leads to an effective additional radiative forcing orders of magnitude larger due to greenhouse gases emissions. In addition, second order *feedback* effects also play a role. For example, increased temperatures due to increased concentration of a certain climate gas may lead again to increased melting of polar ice caps which again increases the temperature even further due to less direct reflection of the incoming radiation.

1.5.3 The Carbon Cycle and Enhanced Global Warming

In closing the present section, let us consider the global flow of carbon (mostly as CO_2 in the atmosphere) presented in Fig. 1.9.

The oceans represent an enormous reservoir of CO_2, with a total content equal to almost 40 Tt (10^{12} t). The content in the soil is about 2.3 Tt, in Earth vegetation about 0.6 Tt and the atmosphere contains 0.8 Tt. The latter is valid at 360 parts per million (ppm), corresponding to about 0.03 % of the weight of the atmosphere.[9] A natural exchange between soil, vegetation and atmosphere via the photosynthesis is about 0.12 Tt per year. An almost equal amount of about 0.09 Tt CO_2 is exchanged each year with the oceans. The exchange rates depends on the temperature and in equilibrium, the sum of the fluxes vanish. The human contribution now comes as an extra 0.009 Tt annual increase in the contribution to the flow into the atmosphere. Note

[9] In 2014, the concentration of carbon into the atmosphere already exceeds 400 ppm.

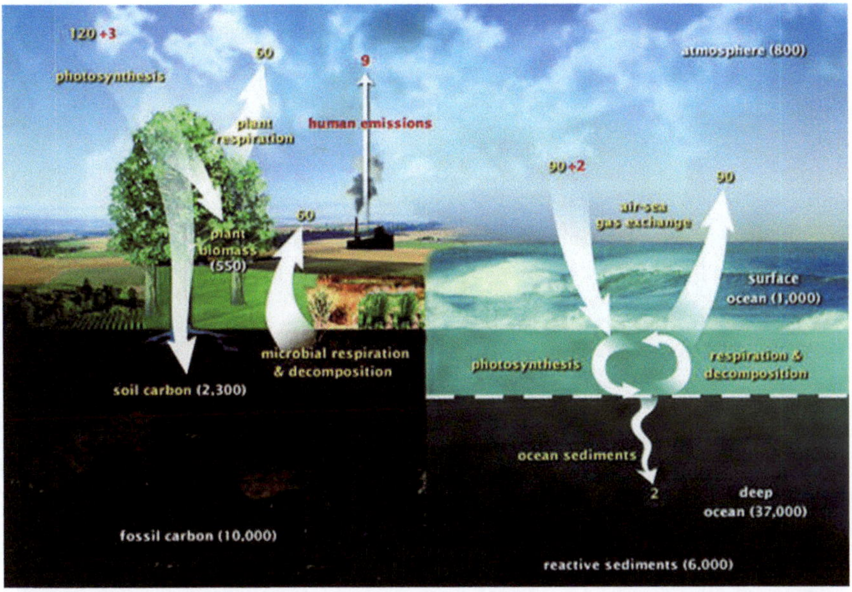

Fig. 1.9 Sketch of the annual amount (in Gtonnes) of carbon exchange globally between land, oceans and atmosphere

the magnitude of these numbers: The human contribution to the total flux upwards is about 4.5 % of the total. The increased fraction of atmospheric CO_2 will at some point again be absorbed by the plant-ocean system. However, since the change of rates are small, increased anthropogenic emission almost exclusively contribute to increased atmospheric CO_2 concentration. Furthermore, an emission of 10 Gt (10^9 t) CO_2 per year imply that it will take about 70 years to double the concentration. Another question is whether we possess enough fossil resources to emit this amount, a topic which will be further discussed in Chap. 2.

It is likely that the change of CO_2 will increase the global temperature above today's normal. How much the temperature will increase is uncertain, but current IPCC numbers forecast an increase between 3 and 5° above preindustrial "normal" level by the end of the century. However, what is "normal" temperature is a difficult issue as well: The Earth's temperature depends on other phenomena such as the Sun's power and cycles, the Earth's orbit around the Sun, ice cap expansion and so on. Thus the average global temperature has varied at all times in the history of our planet, also before humans started burning fossil fuels.

About a thousand years ago, the global temperature was on average comparable with today's temperature, followed by a cold period (the small ice age) lasting up to about 1900. On a longer geological timescale, the climate and the average temperature has in general varied with much larger amplitudes than the changes experienced during the last couple of centuries. Global average temperatures have in certain time spans, for example in the *Carboniferous Period* about 360–300 million years ago, been close to 300 K, more than 10 K above the average temperature today.

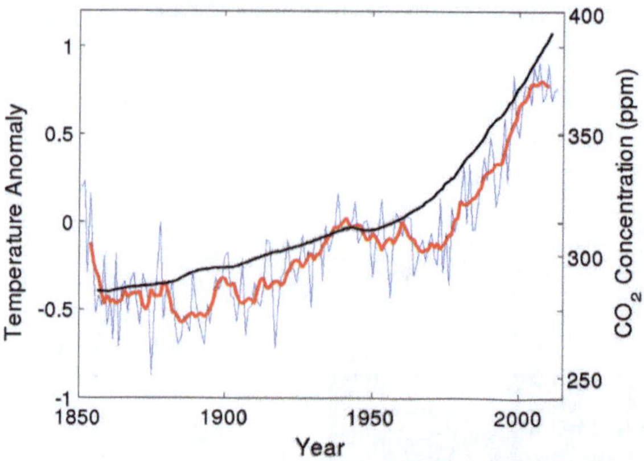

Fig. 1.10 Graphs of averaged global temperature as compared to average CO_2 concentration since 1850. The *blue line* are global average surface temperature data from the Met Office Hadley Centre (http://www.metoffice.gov.uk/hadobs/). CO_2 data are reproduced from a compilation of sources at http://data.giss.nasa.gov/modelforce/ghgases/

The atmospheric CO_2 concentration has in the same period been several thousand ppm, as opposed to today's level of around 400 ppm. On the other extreme, the Earth has experienced several cold periods, the most extreme being the *snowball Earth* period of about 650 million years ago when the entire Earth surface was covered with ice. However, the dramatic different climates of the past took place in a period without humans on the planet. In the most recent 15 years, Fig. 1.10 shows an almost flat average surface temperature even if CO_2 concentrations have monotonically increased during the same period. This is an indication that processes governing climate on Earth are more complex than a linear relationship between CO_2 concentration and temperature, and that global average changes are so slow that it is today in 2014 still too early to state with certainty the global consequences of two centuries of energy supply from fossil fuels. However, it should again be stated that few experts doubt that the climate is about to become warmer. The big question is how warm? And for which consequences?

1.6 Energy Economics

Several approaches are used in practice to evaluate the economics of energy (Bjørndal et al. 2010). Amongst these approaches, the most common are the net present value, the real option and levelized costs of energy. By design, these three approaches usually focus on the **direct** cost of energy, hence how costly it will be for an investor to invest in a project, rather than on the whole cost of a project for society. For example, a wind turbine will only generate power when there is sufficient wind and the system might thus require back-up power to compensate for this *intermittency* issue. In that example, there is a cost of having a wind turbine, cost generally not bared by the investor. This is an **indirect** cost. The sum of the direct and indirect costs gives the full cost of energy.

The net present value approach, the real-option approach and the levelized costs of energy will be reviewed next, followed by a discussion on some of the indirect costs of energy.

1.6.1 Net Present Value Approach

Investments in energy projects are characterized by their irreversability as investment costs will generally be sunk.[10] Investments in wind turbines and solar panels are example of irreversible investments as these will be used until they fail to generate energy, in which case their residual value will be close to zero.

Several techniques are available to evaluate the expected economic outcome of an irreversible investment. The net present value approach (NPV) is the most common of them. It is an approach which can help determine whether a project's financial outcome is expected to be positive or negative. The NPV integrates the initial investment in a project as well as the expected revenues and costs over time, and transforms them in a serie of cash flows adjusted by the time value of money[11] and risk. The basic NPV formula is given hereafter:

$$V_{NPV} = -I + \sum_{t=0}^{T} \frac{C_t}{(1+r)^t} \tag{1.28}$$

In the above formula, I stands for the initial investment, t is the time of the cash flow and T is the duration of the project. The discount rate r is the factor adjusting future cash flows for risk and the time value of money and C_t is the net cash flow. A cash flow in a given year is equal to the expected revenues that year minus the expected costs.

Costs and revenues must be estimated in the most accurate manner in order to lead to useful insights on the financial attractiveness of an investment.

Example Consider the case of a very simple power plant. Investment of a 100 is made in year 0. The plant will generate power in period 1 and 2. Income from sales reaches 80 during the first period and 90 during the second, whereas costs (e.g.: fuel and- maintenance) amount to 15 in both periods.

Cash flows are thus equal to $80 - 15 = 65$ for the first period and $90 - 15 = 75$ for the second period. The investor assumes a discount rate of 7 %.

[10] A sunk cost is a cost that has already been incurred and which cannot be recovered.

[11] The concept of time value of money relies on the rule that a euro today is worth more than a euro tomorrow.

$$V_{NPV} = -100 + \frac{65}{(1+0.07)^1} + \frac{75}{(1+0.07)^2} = 26.25 \tag{1.29}$$

The investor thus expects a positive outcome from investing in this power plant, because the NPV is positive.

A key element of Eq. (1.28) is the discount rate, which is needed to transform future cash flows in order to obtain their present value. Failing to appropriately select the right discount rate has the potential to change the *merit order*[12] between certain projects as it is examplified below.

Example Consider two technologies. The first technology is capital intensive (e.g.: wind project, solar project), which means that most of the cost needs to be bared before the plant starts to produce electricity. In the case of the second technology, most of the cost over the lifetime of the plant is related to the fuel it uses (e.g.: gas-fired power plant, coal power plant). The first technology has capital costs of 900 and operational costs will amount to 10 annually for the 20 years that the plant will be in operation for. The second technology has capital costs of only 200, whereas fuel and operation costs will amount to 60 for the subsequent 30 years. One year is needed to build the plants.

Using a discount rate of 5 %, the present value of the costs of the first project amounts to 1,025.

$$V_{NPV} = 900 + \sum_{t=1}^{20} \frac{10}{(1.05)^t} = 1{,}025 \tag{1.30}$$

The reader can calculate the NPV of the second technology as an exercise.

Figure 1.11 illustrates how a different discount rate impacts the net present value of both projects. It is apparent that capital intensive technologies are less sensitive to variations of the discount rate than other technologies. The reason behind such a big impact on the NPV of non-capital intensive technologies is due to the fact that less and less value is put on future costs when discount rates increase.

[12] The merit order is a way of ranking different energy sources based on their cost of generating power.

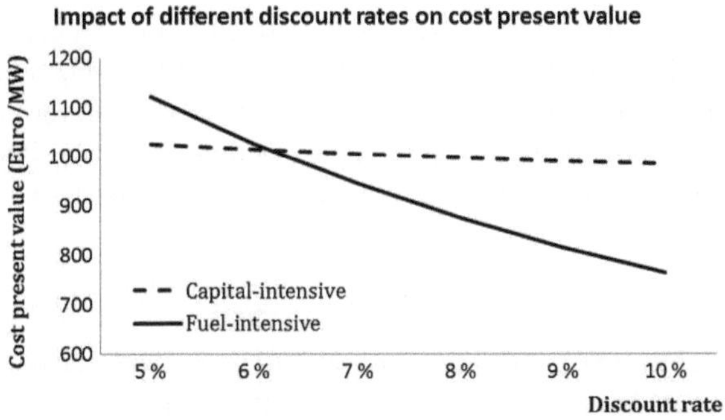

Fig. 1.11 Impact of using different discount rates on two energy technolgies, one of which is capital-intensive whereas the other is not

In this example, everything else being equal, the investor will choose the capital-intensive technology if a low discount rate applies, whereas the other technology becomes more attractive if the discount rate is high.

Figure 1.11 clearly shows that less-capital intensive technologies benefit from a higher discount rate. Though the discount rate impacts differently various technologies, it is important because it puts a value on time preference of money (Branker et al. 2011) and thus allows for the comparison of projects with different economic lives and which costs occur at different times. Keeping these aspects in mind is important when reading reports comparing various energy sources based on costs.

Financial techniques allowing for a proper estimate of the discount rate requires knowledge beyond basic finance. The reader will be able to read about Capital Asset Pricing Model (CAPM), model used in practice to determine the risk-adjusted discount rate, in appropriate corporate finance books.

The example presented in this section is basic. It can however be extended to much more complex forms to integrate stochasticity in the payoff for example. Nevertheless, these cases are beyond the scope of the present book as they also require advanced statistical knowledge.

1.6.2 Real Option Approach

The irreversibility of investments in energy projects associated to the uncertainty in the future value of these projects give rise to an opportunity cost related to the timing of the investment. When facing an investment decision, an investor can either choose to invest now or to wait and keep the possibility of not investing open (Dixit and Pindyck 1994). Waiting can, in some cases, help the investor reduce the uncertainty attached to costs and revenues. Therefore, waiting has a value (known as a real option value) which will drop to zero as soon as the investment is made, because the investor will not be able to benefit from any new information that would potentially have led the investor to adapt her investment decision otherwise. Whenever possible, in the sense that it is cost effective, the value of this option should be integrated in the traditional net present value (NPV value) rule that the investor will apply to decide whether an investment is sound or not.

Example This example provides an easy way to grasp what can be the value of waiting. Let us imagine a wind project with an initial cost of 2,000. At the time the investor is evaluating whether an investment is sound or not, discussions are ongoing at the government level about a new policy instrument which would guarantee a fixed price for each kWh produced from wind farms over a period of twenty years. If this policy instrument is implemented, it would guarantee a payoff of 500 over the next twenty years to our investor, whereas the payoff will only be of 100 otherwise. The investor estimates that there is a probability (p) of 50 % that the policy instrument will be implemented. A discount ratio of 10 % is hypothetized.

If payoffs are high, the net present value of the project is:

$$V_{NPV} = -2{,}000 + \sum_{t=1}^{20} \frac{500}{(1.1)^t} = 2{,}557 \tag{1.31}$$

If payoffs are low, the net present value of the project is equal to:

$$V_{NPV} = -2{,}000 + \sum_{t=1}^{20} \frac{100}{(1.1)^t} = -1{,}148 \tag{1.32}$$

Traditional finance would encourage us to handle uncertainty by using expected payoffs equal to $E(R_t) = (p * 500) + [(1 - p) * 100] = 300$ in which case, the NPV amounts to:

$$V_{NPV} = -2{,}000 + \sum_{t=1}^{20} \frac{300}{(1.1)^t} = 554 \tag{1.33}$$

Based on the NPV, the investor should invest even though she might end up losing 1,148 with a probability of 50 %! If the investor has the option to wait, she might be better off waiting to know whether the policy instrument is being implemented. The traditional NPV can be modified such as to integrate the value of waiting an extra year.

$$V_{NPV} = (0.5)\left[\frac{-2{,}000}{1.1} + \sum_{t=2}^{21} \frac{500}{(1.1)^t}\right] = 1{,}026 \tag{1.34}$$

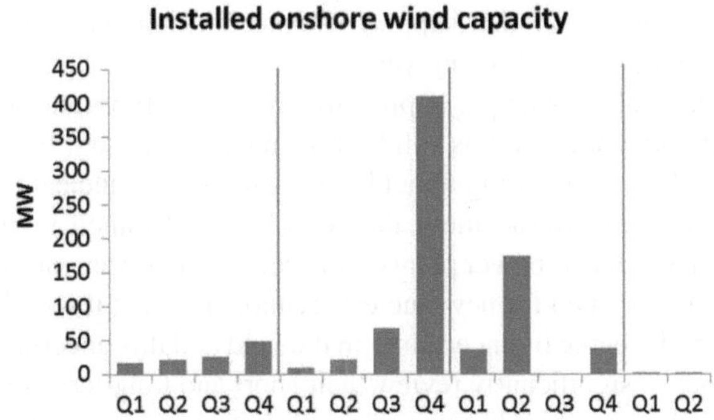

Fig. 1.12 Quarterly installed onshore wind capacity in Denmark (Danish Energy Agency 2013)

Equation (1.34) integrates the 50 % probability that the policy instrument is introduced in one year. As the investor decides to wait, her project will generate payoff only from year 2 (she only invests if the policy instrument is implemented) and until the end of year 21, when the wind turbines will be scrapped. The investor is much better off if she waits an extra year. The value of waiting amounts to $1{,}026 - 554 = 472$ in this example. It means that the investor would be ready to pay as much as 472 to know today if the policy instrument will be implemented or not; it is the real option value.

The incentive to wait was provided in the example above by the possible implementation of a new policy instrument. Other elements have the potential to encourage investors to wait for more certainty before investing, e.g. the expectation of the release of a new wind turbine which would be cheaper and more efficient. One may wonder at this stage why, in fact, there is money invested in renewable energy projects if waiting could bring bigger payoffs. Several cases exist that motivates project developers to push their project forward. The most obvious case is when the value of waiting becomes negative due to a reduction of an existing support scheme. Such case occured in Denmark at the end of 2002. Until december that year, a Feed-in Tariff (FiT)[13] of 0.43 Danish kroner/kWh was in place. This FiT was then reajusted to 0.33 Danish kroner/kWh in January 2003.

The evolution of the quarterly installed onshore wind capacity in Denmark shown on Fig. 1.12 indicates that the value of waiting in December 2002 was negative. Over 250 wind turbines were commissioned during the last two months of 2002, just before the change in FiT which would have made these wind turbines less attractive financially. After the regulation change, some investment took place in 2003, which

[13] A Feed-in Tariff means that a fixed price is guaranteed for each kWh fed into the electricity grid for the first few years of a project. Such measures are implemented to support the deployment of a technology which would not be able to compete otherwise.

coincided with the end of another support scheme. Virtually no wind turbines were commissionned during the following year.

The real option value principle applies to any energy types, not only to renewables. The regulatory uncertainty is valid for the nuclear industry, for example. Prior to Fukushima, talks were ongoing about building three new nuclear power plants in Switzerland. Shortly after the accident, the Swiss Federal Council advised against the construction of new nuclear power plants. This decision has been modified in August 2011 to leave the door open for new nuclear technologies, but it was definitely sufficient to bring a high degree of uncertainty in the field and this uncertainty led several energy companies to significantly review their short and long-term strategies. In the coal industry, fuel cost is highly uncertain and the absence of clear environmental legislations in various countries is slowing down the construction of new coal power plants.

There are several types of real options (Fernandes et al. 2011). The type presented in this section is known as a 'defer option', i.e. the value of waiting. Another relevant real option in the energy field is the 'growth option', which is related to the possibility to expand an energy project in the future. The value of this option is particularly high for energy projects that can be built in modules, since such projects can more flexibly adapt to changing market conditions. For example, if a power plant is built to match the growing energy demand, it is possible that future growth prospects are highly uncertain. In this case, the flexibility value offered by a plant to which subsequent units can be added at will, needs to be taken into account if compared to the construction of a larger, unflexible plant. The former type allows the investor to retain some of the real option value. And if bigger plants usually lead to better economies of scale compared to more flexible plant, the investor will favor flexibility under the real option approach until the value of the economies of scales exceeds the value attached to this flexibility. The NPV approach typically ignores flexibility.

1.7 Levelized Costs of Energy (LCOE)

The NPV- and the Real Option approaches are both relevant to investors since they are tools providing insights on whether a project is expected to be profitable or not and on when and how it is best to invest to maximize the expected returns. Another approach known as the levelized cost of energy (LCOE) is used by investors, researchers and governments to compare technologies relying on different physical principles, use different fuels and which can have dissimilar economic plant lifes. The LCOE has been designed with the idea of allowing for comparisons between energy sources on a unit cost basis over the lifetime of different energy technologies/projects. Criticisms of this approach are that it does not account for specific market or technology risks (e.g.; uncertainty in future fuel costs) (OECD/Nuclear Energy Agency 2010) nor does it account for elements such as intermittency and the need for back-up power. Yet, it remains the best method available to compare various energy technologies.

Three distinct elements, namely capital costs, operation and maintenance costs and fuel costs, are evaluated under the LCOE approach. These are separately reviewed before being combined to create an equation useful to determine the levelized costs of energy. Three examples will then be described to provide the reader with a feel on how the LCOE approach can be applied to concrete cases.

1.7.1 Capital Costs

All energy projects are subject to capital costs[14] because some investment is always needed before a plant can start generating energy. This initial investment can be relatively small (e.g.: a micro wind turbine installed on a rooftop) to very large in absolute values (e.g.: large nuclear power plant or hydropower dams).

In the complete LCOE formula, capital costs are labelled c_p, which stands for the cost of the plant. It is measured in monetary units by unit of installed capacity (e.g.: Euro/kW or Euro/MW). A concept named capacity factor is needed in order to translate this absolute number into a unit cost basis (e.g.: Euro/MW**h**). This capacity factor f is the power produced over a period of time divided by the power that may have been produced if the plant was running 100 % of the time over the time period considered. The capacity factors turns around 30 % for intermittent electricity generating plants (e.g.: wind, solar) to above 90 % for some nuclear power plants.

The typical period of time considered in a LCOE approach is one year, which is equivalent to 8,760 h (labelled H). The capital cost per unit of energy generated over a year of operation can now almost be estimated using the previous information.

It remains to take into consideration that power plants have economic plant lives of several decades, which means that a last twist needs to be taken into account in order to properly estimate the capital cost per unit of energy generated. This twist is known as the capital recovery factor R, which is the share of the plant cost that the income must cover over each year of operation such as to balance out the whole project at the end of the plant life. In more formal terms, it converts a serie of equal annual payments over a set length of time into a present value. Two terms are needed to compute the capital recovery factor: the discount rate r and the economic plant life T.

$$\text{Capital costs} = \left[\frac{R \cdot c_p}{H \cdot f}\right] \tag{1.35}$$

$$\text{where } R = \frac{r \cdot (1+r)^T}{(1+r)^T - 1}$$

[14] Also referred to as overnight costs in some studies.

1.7.2 Operation and Maintenance Costs

A second category of costs is relevant to all energy technologies. These are the operation and maintenance costs (O&M), which can be fixed or variable. For practicallity purposes, both variable and fixed O&M costs are grouped together under the label c_o in the formula below. It is obvious that separating them would give more precise LCOE, however the data available is seldom detailed enough. If we assume that we have access to a given number of O&M costs valid for the first year of operation of a power plant, it is possible to convert this number to a unit cost basis by using the terms H (hours per year) and f (capacity factor). A twist is also possible in order to integrate an increase of the O&M costs as the plant ages (higher chance of major breakdown, need for more maintenance), which is known as the levelization factor (l). This factor depends on the discount rate r and on an escalation rate e measuring by how much the O&M costs are expected to increase annually.

$$\text{O\&M costs} = \left[l \cdot \left(\frac{c_o}{H \cdot f} \right) \right] \tag{1.36}$$

$$\text{where } l = \frac{r \cdot (1+r)^T}{(1+r)^T - 1} \cdot \frac{(1+e)}{(r-e)} \cdot \left[1 - \left(\frac{1+e}{1+r} \right)^T \right]$$

1.7.3 Fuel Costs

A third category of costs is related to the fuel. The cost of fuel can be nonexistent in the case of renewable energy technology or represent a large portion of the cost per unit of energy generated (e.g.: fossil-based power plants). The cost of fuel is labelled c_f. When the fuel costs for a year are known, these can be transformed into a unit cost basis by integrating a levelization factor, the capacity factor and the number of hours in a year.

$$\text{Fuel costs} = \left[l \cdot \left(\frac{c_f}{H \cdot f} \right) \right] \tag{1.37}$$

1.7.4 Full LCOE Formula

The complete version of the LCOE approach (Roth and Ambs 2004) is presented below (Table 1.8).

$$C_{LCOE} = \left[\frac{R \cdot c_p}{H \cdot f} \right] + \left[l \cdot \left(\frac{c_o}{H \cdot f} \right) \right] + \left[l \cdot \left(\frac{c_f}{H \cdot f} \right) \right] \tag{1.38}$$

Table 1.8 Nomenclature of the various components of the LCOE approach

Nomenclature			
c_f	Fuel cost (in Euro/MW)	H	Hours per year
c_o	O&M cost (in Euro/MW)	l	Levelization factor
c_p	Plant cost (in Euro/MW)	r	Discount rate (in %)
e	Escalation rate (in %)	R	Capital recovery factor (in%)
f	Capacity factor (in %)	T	Plant life (in years)

$$R = \frac{r \cdot (1+r)^T}{(1+r)^T - 1}$$

$$l = \frac{r \cdot (1+r)^T}{(1+r)^T - 1} \cdot \frac{(1+e)}{(r-e)} \cdot \left[1 - \left(\frac{1+e}{1+r}\right)^T\right]$$

The LCOE approach has to be adapted to the data available and to the purpose of the study. For example, an investor relying on the LCOE approach will generally focus only on the direct costs of a project, whereas a policy maker may want to expand the approach and include *externalities*[15] in the calculations.

It goes without saying that a different escalation rate can be used for O&M and fuel costs, especially if it is believed that these costs will increase at a different pace. However, predicting the escalation rate of any fossil fuel with accuracy is challenging, if not impossible. For that reason, most existing studies choose to use only one escalation rate.

Example A The levelized costs of a fictive wind power plant (LCWP) are calculated here using the LCOE approach. The following parameters have been imagined by combining existing estimates (Krohn et al. 2009) and plausible assumptions from the authors:

- Plant cost: 1,227 Euro/kW (Krohn et al. 2009)
- Capacity factor: 25 %
- O&M costs: 0.012 Euro/kWh (Krohn et al. 2009)
- Discount rate: 10 %
- Escalation rate: 1 %
- Plant life: 25 years

The capital recovery factor can conveniently be computed with the data available and amounts to 11 %. This recovery factor means that it is necessary to generate a yearly income equivalent to 11 % of the plant cost to cover the sum

[15] Cost or benefit not transmitted through price.

of all fixed charges for one year of operation. The plant costs include, among others, the costs associated to the turbines, the cost of connecting the turbines to the grid and the foundations. The capacity factor of 25 % means that the wind turbines are expected to generate 25 % of their maximum theoretical power output throughout a year of operation. O&M costs integrate the fixed and the variable part and notably include administrative costs, the land rent, insurances, service and spare parts. Wind power does not require fuel and thus, the fuel cost is set to 0. The O&M escalation rate is set here to 1 % meaning that O&M costs are expected to increase by 1 % every year as the wind turbines get older and requires more maintenance work. Given the data available, the LCWP can be estimated:

$$C_{LCWP} = \left[\frac{R \cdot c_p}{H \cdot f}\right] + [l \cdot c_o] = 0.075\,\frac{\text{Euro}}{\text{kWh}} \tag{1.39}$$

And the result of Eq. (1.39) indicates that the LCWP amounts to 7.5 Eurocents/kWh. This result means that if the producer can sell her expected production at an average price exceeding this levelized cost, she can expect to generate a return over investment.

Estimating the levelized costs of a fossil-based power plant is marginally more challenging since some key physical concepts need to be included when computing the costs. The cases of electricity generation from natural gas and coal are detailed thereafter and these should help the reader to properly estimate the costs of processes relying on any fuel.

Example B The basic process behind a natural gas-fired power plant is that heat is released when natural gas is burned. This heat can be used to heat up water, which will boil and create steam. This steam can in turn be used to spin a steam turbine and generate electricity. The first element of interest here is that natural gas is needed in the process, the cost of which is often given in USD/MMBtu (British thermal unit) or Euro/MMBtu. Transformation will therefore be needed to translate Euro/MMBtu to Euro/MWh, which is the unit used to measure the LCOE. The second element of interest is that not all power plants manage to transform natural gas with the same efficiency, meaning that different power plants will require various quantities of fuel to generate a given amount of electricity. These differences are related to the thermal efficiency of a process, which is:

$$\eta_{th} = \frac{W_{out}}{Q_{in}} \tag{1.40}$$

The thermal efficiency is therefore a measure of how much work can be obtained by unit of input. The higher the efficiency, the lower the quantities of fuel needed per unit of energy generated. For illustration purposes, we assume that the cost of natural gas is equal to 4.50 Euro/MMBtu. Knowing that 1 Btu = 1,055 J:

$$4.50 \frac{\text{Euro}}{\text{MMBtu}} = X \cdot 1.05506 \frac{\text{GJ}}{\text{MMBtu}} \tag{1.41}$$

Where X = 4.27 Euro/GJ. As mentionned above, the work that can be obtained by unit of input is limited by the thermal efficiency of the plant. The following relationship can be used to estimate how much fuel is needed for our 'inefficient' plant to generate as much heat as a theoretical plant that would reach the physically impossible mark of 100 % efficiency. Let's assume a thermal efficiency of 57 %.

$$\frac{1}{\eta_{th}} \cdot 3.6 \frac{\text{GJ}}{\text{MWh}} = 6.3 \frac{\text{GJ}}{\text{MWh}} \tag{1.42}$$

Which indicates that 6.3 GJ worth of fuel will be needed to generate one MWh of electricity. Equations (1.41) and (1.42) can be combined to obtain the wanted value:

$$6.3 \frac{\text{GJ}}{\text{MWh}} \cdot 4.27L \frac{\text{Euro}}{\text{GJ}} = 26.94 \frac{\text{Euro}}{\text{MWh}} \tag{1.43}$$

For this particular plant, the fuel needed to generate one MWh of electricity costs 26.94 Euro or 2.7 Eurocents/kWh. The LCOE formula can then be used to estimate the LCOE of that plant.

Example C A similar approach can be applied to coal, for which prices are generally given in USD/tonne or Euro/tonne (other prices such as USD/short ton are also used in practice). It will be made clear in the section on coal that this fuel can be divided in several classes depending of the energy content of the fuel. Several elements are therefore needed in order to translate the price of a coal product into a unit useful to calculate the LCOE of coal. These elements are:

- The price per tonne
- The mass-to-heat conversion factor (which is the element allowing for the incorporation of the quality of the fuel) and
- The thermal efficiency of the plant

Assumptions for a power plant using coal are listed below:

- The price per tonne = 60 Euro
- The mass-to-heat conversion factor = κ = 22 GJ/tonne
- The thermal efficiency of the plant = 39 %

Similar to what has been done for natural gas, the following equations can be combined:

$$60\frac{\text{Euro}}{\text{tonne}} \cdot \frac{1}{\kappa\frac{\text{GJ}}{\text{tonne}}} = 2.72\frac{\text{Euro}}{\text{GJ}} \tag{1.44}$$

$$\frac{1}{\eta_{th}} \cdot 3.6\frac{\text{GJ}}{\text{MWh}} = 9.23\frac{\text{GJ}}{\text{MWh}} \tag{1.45}$$

$$9.23\frac{\text{GJ}}{\text{MWh}} \cdot 2.72\frac{\text{Euro}}{\text{GJ}} = 25.11\frac{\text{Euro}}{\text{MWh}} \tag{1.46}$$

For this particular plant, the fuel needed to generate one MWh of electricity thus costs 25.11 Euro.

In the following chapters, the LCOE are calculated for all major energy sources. Unless stated otherwise, the monetary values are given in ^{2008}Euro and a unified discount rate of 10 % is utilized. Data on overnight costs, O&M costs, electrical conversion efficiencies and load factors are taken from a joint report by the International Energy Agency and the Nuclear Energy Agency (OECD/Nuclear Energy Agency 2010), whereas the remaining assumptions (e.g.: fuel cost) are those of the authors.

Additional steps to convert from any currency to ^{2008}Euro will be taken to allow for easy comparisons throughout the book.

1.8 Other Relevant Economical Aspects

The previous sections elaborated on approaches useful to evaluate the expected financial outcome of an investment and to compare technologies' costs together. In this section, we will start by discussing how to estimate the escalation rate of fuel, followed by an introduction to an important category of costs which has been ignored so far: the indirect cost of energy.

1.8.1 Introduction to Resource Economics

Coal, natural gas and oil are finite resources and consequently non-renewable. It implies that once a resource is consumed, it is not possible to recover any of it. Therefore, the owner of fossil fuel reserves must not only consider how much profit can be made by extracting the resource today, but also if it would not be wiser to keep the resource in the ground and sell it at a later stage when economical conditions are more favorable. In exploring what is the optimal exploitation course of an exhaustible resource, Hotelling (1931) defined what is known as the 'fundamental principle' of resources extraction. Assuming perfect market conditions and that prices will increase as resources get depleted, the principle states that in order to justify the extraction of a finite resource, the price received by the resource owner after the costs of extraction and the costs of placing the resource on the market have been paid for, must rise at the same rate as the market rate of interests. This concept is theoretically easy to understand, because if the market rate of interest rises faster than the net price received by the owner of the resource, the rational profit maximizer should deplete her resources as quickly as possible and invest the profit elsewhere. In the case where the market rate of interest rises slower than the net market price of the resource, the rational profit maximizer should keep her reserves in the ground. The equilibrium condition indicates that the cost of a resource in real terms should rise exponentially over time, given that the resource is exhaustible. According to Hotelling (1931), real market prices for non renewable natural resources should evolve based on market conditions.

Slade (1982) adapted this theory in 1982 such as to allow for a decline in extraction costs. She recognized that a decline in extraction costs can outweight the increase in rent (yield from the resource deposit) until a certain point where such costs are outweighted by scarcity costs, leading to a U-shaped trend for the extraction cost of a non-renewable resource.

Both these models are simple (as Slade admitted herself) and recent research tries to create more elaborated models that would suit reality better and eventually allow for improved predictions on future resource prices. Such an example is Zacklan et al. (2011), where the authors establish that resource prices are characterised by stationary and non-stationary periods (stationary means that an exogenous shock affecting a time-serie is quickly absorbed whereas the effect of such shock on a non-stationary period does not die out) . In the former case, market conditions (such as described by Hotelling (1931)) explain fairly well the evolution of a resource price. In non-stationary periods, exogenous shocks impact the prices and traditional models do no longer apply. Such structural break have at least occured in 1945 and 1973. Even though such approach brings light on the evolution of a resource's price, it requires long time series before a change from a stationary to a non-stationary trend can be established. It has therefore limited power in predicting prices in the near future.

Hotelling (1931) and Slade (1982) developped their models at a time when climate change was not a concern. The latter may well distort these models as expectations of future environmental regulations may lead a resource owner to extract the resource as fast as possible before such regulation prevents her to do so or renders it less profitable. Future environmental regulations are not the only factor affecting the extraction levels. Fluctuations in extraction cost and change in interest rate (Kula 1992) will also impact significantly the extraction rate.

Gas-fired and coal-fueled power plant owners, as well as energy-intensive industries based on fossil fuels are following closely the evolution of the various fossil fuel prices as these will impact the economical competitiveness of many energy sources. The respective sections on natural gas and coal will demonstrate that the share of the total electricity cost represented by the fuel cost is significant and defavorable prices can make a project unattractive financially. Investors therefore need to properly estimate the fuel prices in the future in order to get correct levelized costs for their electricity and/or heat they will be producing. As this short introduction to resource economics illustrates, proper future fuel prices estimates are difficult, if not impossible. In this respect, investors usually use a steadily increasing fuel price over the lifetime of the powerplant. This approach reflects the 'fundamental principle' from Hotelling and the increasing part of Slade's U-shaped part of the cost curve.

1.8.2 Indirect Costs of Energy

Let us consider a few distinct cases:

- Wind power, solar power or small-scale hydro are intermittent energy generating technologies, meaning that these will only generate energy when it is windy, sunny or when the water flow is sufficient. Since these cannot be obtained at will, other technologies may be needed to match demand and supply of power. This relates to the concept of intermittency.
- Nuclear power plants and other thermal plants are regularly stopped for inspection and maintenance purposes. During these periods, other technologies need to replace the plants under maintenance, resulting in a cost increase at the system level.
- Tapping into the renewable energy potential in developped countries may require the extention or the strenghtening of the existing grid, often at great costs. Solar PV optimally located in developing countries can avoid the need for large infrastructure as a very local grid can suffice to provide residents with electricity.
- Integrating PV into the grid usually increases the overall system cost because intermittency has to be managed, the grid may have to be strenghtened, etc. But in cases of heat waves as in France in 2003 or more recently in the United States, the efficiency of thermal plants decreases and in extreme cases, some thermal plants have to be shut down, putting pressure on the system. In such cases, electricity generation from solar PV is at its maximum, thus 'easing' the system.
- Finding a new environmentally-friendly fuel that seems competitive with gasoline and diesel on a direct cost basis might become totally unattractive when indirect costs of adapting the infrastructure and changing existing cars are taken into account. This last phenomenon refers to the concept of path dependence.

These five cases illustrate that direct energy costs may not be representative of the full cost of integrating a technology in a given system. Ideally, indirect costs would be added to the direct cost of energy in order to get the cost of an energy source at a system level. This is however challenging. The difficulty in estimating the complete cost of a specific energy technology stems from the fact that this cost

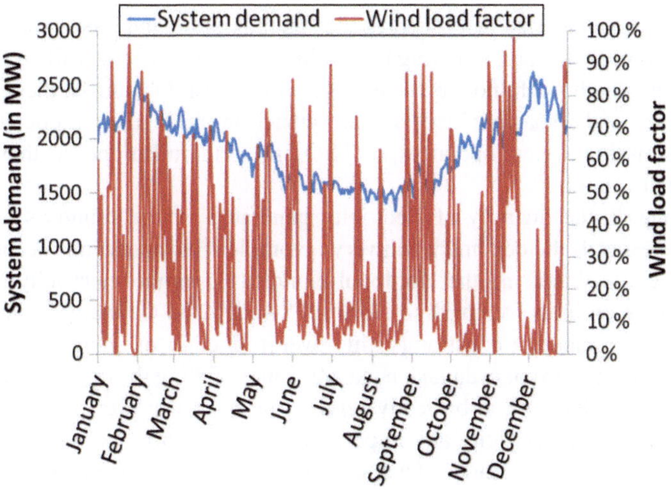

Fig. 1.13 Variability of an intermittent energy source output and total electricity demand for Western Norway in 2012

will depend on the system where this technology operates. For example, the level of market penetration of each technology will impact the indirect costs of integrating one specific intermittent technologies. Therefore, managing one wind turbine into a large grid is simple, whereas a 20 % penetration level will require a much better infrastructure. The reader must nevertheless remember that when it is stated that an energy technology costs X, it is only part of the picture.

1.8.2.1 Intermittency Cost and Dispatchability

The most known issue about intermittent energy generating sources is precisely their intermittency. This concept applies to all technologies, which production depend on external factors (e.g.; wind, solar radiation, water flow). An energy source such as wind-, solar- or small scale hydro power will be subject to seasonal, daily and hourly variations. However, the scale of the variations depend on the energy source considered. If electricity generated from wind can drop from maximum to zero within a matter of minutes, solar panels will generate electricity when there is daylight and none at night. In addition, these intermittent sources of energy are non dispatchable, implying that these will generate power when they can, as opposed to when we need it. Figure 1.13 illustrates the concepts of intermittency and dispatchability.

Figure 1.13 shows the total daily electricity consumption in Western Norway (in blue). A seasonal pattern is clearly visible as consumption is higher in winter than in summer. The daily wind load factor (electricity effectively generated divided by the rated power) is represented in red and its behavior is anarchic with no precise pattern. The concept of dispatchability is very clear as it is shown on the figure that there are times when the wind load factor approaches its maximum and times when the wind load factor is close to zero, implying that an intermittent generator will be producing when it can, which might not be when it is needed most. Obviously, these issues of intermittency and dispatchability have an impact on the regulating operators' task (i.e.; transmission system operators, or TSO, in Northern Europe) as they manage the electricity supply.

In Denmark, the quantity of electricity generated by the country's wind park sometimes exceeds the demand when very favorable wind conditions occur. In such cases, the country is using the Nordpool market (the largest electricity market in the world) which links Denmark to Norway, Sweden and Germany. However, due to transmission capacity constraints, this overproduction of energy can drive the electricity price to zero or even take negative values. Taking the opposite case, when wind is scarce, Denmark is basically relying on its coal power plants to provide its inhabitants with electricity and imports hydropower from Norway and Sweden to balance its electricity market. Overall, this means that Denmark stores its wind in the form of water in Norway and Sweden (Green and Vasilakos 2012) for which Denmark pays a price equivalent to the difference in price between import and export of electricity.

The precise impact on the total cost of the system of integrating intermittent energy sources in one particular system is unique to each system, because the characteristics of each system are indeed unique. Differences in balancing capacity, consumption patterns, transmission grid capacity (IEA 2011), energy mix, quantity of intermittent power and their location will affect the cost of integrating these technologies.

Most of the renewable energy types currently under development are subject to this intermittency and dispatchability issue. Tidal-, wave-, concentrated solar-, photovoltaic-, small hydro- and wind power all fall under this category. Yet, the output of some technologies, e.g.: wind power, is generally more variable than the output of other technologies, e.g.: concentrated solar power. Other renewable energy technologies do not belong to this category of energy, notably power from biomass and hydropower. Storage can also mitigate the intermittency issue. For example, the intermittency issue from wind can be suppressed by using the concept of pumping storage, i.e.; using the energy generated from the wind turbine to pump up water in reservoirs, where this energy will be stored as potential energy and run through hydropower turbines when power is needed.

Outside the electricity sector, the quasi-totality of the energy used is, for obvious reasons, dispatchable. When powering a car from renewable resources, it is better if it can be used at will. Thus, biofuels, hydrogen, batteries all provide reliable and dispatchable power.

1.8.2.2 Techno-Spread and Geographical Dispersion

The reader might think at this stage that integrating intermittent power in the system means that there is a need to build an equivalent amount of backup capacity in order to compensate for the failure of the intermittent sources. It can be true in a small system

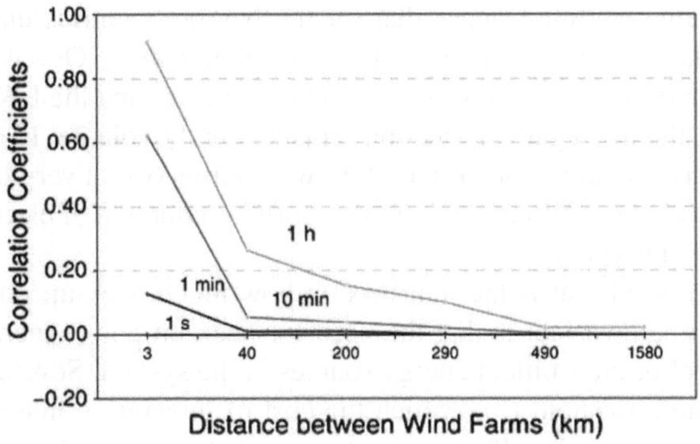

Fig. 1.14 The correlation coefficient between wind from various location decrease as distance increase

where there is one type of intermittent energy source only. Such an example would be the island of Utsira off the coast of Norway. This island gets its electricity from wind and when there is electricity in excess, it is transformed into hydrogen which is used to supply the market when wind is scarce. In bigger systems, the statement that there is a need for a one to one ratio between intermittent capacity and backup power is incorrect. First, combining multiple intermittent energy technologies can reduce the need for backup. Second, geographical dispersion also reduces this need (Fig. 1.14).

Natural elements such as wind, solar radiations, wave and tides are not all correlated with each other. On the contrary, they even tend to be complementary. Such type of diversification can be refered to as 'techno-spread' (OECD/Nuclear Energy Agency 2010). Therefore, the risk of not having any production from intermittent energy sources decreases as more intermittent energy types are included in the market.

Geographical dispersion also reduces the need for backup capacity since, for example, wind correlation between two different locations decreases when the distance between these locations increases. The result of one of these studies is in Fig 1.14.

By connecting the intermittent energy sources together and given that there is sufficient distance between them, the geographical dispersion factor will help reduce the need for backup power. By looking at one wind turbine, the output can be maximal at some stage and drop to zero over a short period of time. Looking at the output of many wind turbines spread over a large area and interconnected shows that the aggregated output almost never reaches zero or the maximum but follows a smoothened pattern.

Attempts have been made to precisely estimate how much back-up power is effectively needed to compensate for the intermittency factor. Again, estimates have to be done for each system individually as several factors will have an influence on the result. For Finland, one study (Holttinen 2004) estimates that for a penetration level of up to ten percent of wind energy, only two percent of back-up power is needed. The same study advances that for the Nordic countries united under the Nordpool market, one percent could suffice. In another study, Oswald et al. (2008), the authors found that although spreading wind farms around the UK smoothes the flow of power, the aggregate production remains highly volatile. In addition, large weather systems can affect the output of all wind farms over a very large area for a prolonged period of time, thus leading to no or full production of electricity from all wind farms in some events.

It is to be noticed that if the numbers on how much back-up power is needed are uncertain, a known fact is that the need for back-up power increases with the penetration level of intermittent energy sources in the system. Several authors have tried to estimate a treshold over which the cost of integrating intermittent energy sources would be excessive. Such estimates vary between twenty to thirty percent, however, such penetration levels remain to be seen in large systems.

In addition, efforts towards managing demand can relieve the system. The generalisation of a smart grid where demand would react more to prices and in cases where demand is controlled by the system operator (such as switching off heaters in houses from a distance) can lower the demand and reduce the need for back-up power. Other less obvious solutions are possible to facilitate the balancing of the market. Achieving more flexible thermal generation is one such example. Coal-based thermal units are slow to start, whereas gas-fired thermal units can adjust much more easily. Thus, replacing coal by natural gas increases the flexibility of a system.

Taking all the factors mentionned in the past two sections and more into account, it is estimated that the cost of balancing the system due to the presence of intermittent energy sources is between 0.7 Eurocent/kWh to 4.7 Eurocent/kWh (in 2008 Euro) at 20 % penetration level of intermittent energy sources (IEA 2011), with higher balancing costs in the United Kingdom as its electricity system is relatively unflexible and lower costs in Norway and Sweden.

1.8.2.3 Path Dependence

An interesting feature explaining why certain technologies struggle to penetrate the market is path dependence. This concept can also be used to explain differences between countries in terms of their energy mix as it includes history on top of the market conditions. The most famous example of path dependence is the QWERTY keyboard (David 1997).

The QWERTY keyboard is still being used today even though it has been proven that it is not the most efficient design. In fact, it was made such as to slow down (!) the person typing on a keyboard at a time when typewritting machine could not handle quick typing. However, even though the technology has improved tremendously in the past 30 years, the QWERTY design is still in use today. Path dependence thus means that the current development of technology relies not only on the current market conditions (technology available, cost, etc.) but also on historical conditions.

The concept of path dependence does also apply to energy. For example, Germany and France can be considered as sister countries in the sense that they are relatively similar in terms of climate and economical development. Yet, the energy mix of Germany and the energy mix of France are very different. Germany relies mostly on coal (24 % of total energy consumption in 2008) and less on nuclear (11.5 %) and renewables (8 %), whereas France relied heavily on nuclear (42 % of total energy consumed) and little on coal (5 %) and renewables (7.5 %). Path dependence can explain such difference as Germany relied on coal in its early industry, whereas France did not. Therefore, when Germany needed more power, it was natural that the country turned towards coal as they had the knowledge, experience of it and a well developped coal industry. Path dependence can also explain why a number of technologies struggle to penetrate the market. For example, if a country's infrastructure can perfectly cope with gasoline and diesel powered cars, this infrastructure might need to be replaced partly or completely if a new type of fuel relying on a different infrastructure enters the market. As it is costly to replace the infrastructure, it is logical that substitute to gasoline and diesel that can be distributed using the existing infrastructure will be prefered compared to a possibly cheaper, more environmental and better fuel.

1.9 Exercises

1. **Units**

   Take a look at Table 1.2 and calculate the power per capita in 2008 in India and in USA i kWh/day.

2. **Mechanical energy—energy conservation**

   (a) Water is stored in a reservoir 500 m above sea level. Assume it can flow without friction in a tube towards the sea. Calculate the velocity of the water when arriving at the sea. Discuss how friction modifies the calculation.
   (b) Prove that the power from the water at the end of the tube is $P = \frac{1}{2}\rho A v^3$ where $\rho = 1{,}000\text{kg/m}^3$ is the density of water and A is the cross sectional area of the tube.
   (c) Assume the tube to have a cross sectional area A = 1.5m^2. Calculate the frictionless power of the water flow.
   (d) Assume the turbine to convert 80 % of the theoretical limit of available power in (c) into electrical power. Discuss where the rest of the power is "lost".
   (e) How many people in a standard western society can be supported by the power plant?

3. **Matter and Photons**

   (a) An excited atom "falls down" to its ground state. This is accompanied by the emission of a photon with energy, $E_{photon} = 4\text{eV}$. Find the frequency and the wavelength of the photon. Does it correspond to visible light? ($1\text{eV} = 1.6 \times 10^{-19}\text{J}$)
   (b) Take a look at Fig. 1.8 and find the typical photon energies which CO_2 emits and absorbs radiation from.

4. **Radiation power fluxes—Earth**

   Based on Fig. 1.8 what is "likely" to happen with the global temperature if

   (a) ... the concentration of CO_2 increases ?
   (b) .... the concentration of H_2O increases ?

5. **On the amount of CO_2 in the atmosphere.**

   (a) From Fig. 1.9 we see that the atmosphere contains about 800 Gtons of carbon. Given that the atomic weight of carbon is 12 u and the atomic weight of oxygen is 16 u, calculate the mass of atmospheric CO_2. Note: 1 u is equal to 1.66×10^{-27}kg.
   (b) Calculate the mass of the entire atmosphere from the following simplified model: A constant density atmosphere with $\rho = 1.2\ \text{kg/m}^3$ which extends up to 10 km above the Earth surface. The radius of the Earth is given in Table 1.7. Compare your answer with the correct number, $m = 5.1 \times 10^{18}$ kg
   (c) The weight fraction of CO_2 in the atmosphere is about 0.06 %. From your result in b), calculate the total mass of atmospheric CO_2 and compare it with your result in a)
   (d) With a annual 30 Gt global emission of CO_2, calculate the time it will take to double the atmospheric concentration.

6. **Energy consumption of petrol versus electric car.**

   Here are approximate data for two comparable (medium size) cars, e.g. a Peugeot 208 (petrol) versus Nissan Leaf (electric). Petrol car: m = 1,000 kg, energy consumption: 0.5 l per 10 km, price 230,000 Norwegian Kroner (NOK). Electric car: m = 1,500 kg (500 kg is batteries), 24 kWh battery capacity gives a driving range of about 100 km on average in Norway. Energy needed to produce the car can be assumed to be 30 MJ/kg for both, and additionally 90 MJ/kg for the batteries.

(a) Assume both cars is driven 200,000 km during a 10 year life cycle. What is the total energy consumption of each car?
(b) Given an (assumed) average price of petrol equal to 15 kr per litre, and a price of electricity equal to NOK 0.5 per kWh, and no additional expenses on the two cars during the 10 year period. What is the total cost per car (to buy and to drive) ?
(c) Give reasons (pro and contra) for having lower taxes on electric cars in a climate perspective.

7. **NPV and real option**

Consider a possible investment in a tidal power project. The plant has a cost of NOK 1.8 million and the owner can either invest in the beginning of year 0 or in the beginning of year 1. If the owner chooses to invest in year 0, the tidal installation will be scrapped at the end of year 20. In year 0, the tariff for energy from tides is set at 200 for a years' worth of energy production (NOK 200,000 in revenues). There exists regulatory uncertainty for year 1 and you estimate that there is a 20 % probability that the tariff will increase to 300 and 80 % probability that it will decrease to 100 (yearly revenue of NOK 300,000 and 100,000 respectively). In addition, you assume that this tariff will remain at the same level forever. The investment can be made immediately and revenues are generated from the time the investment is made. Use a 5 % discount rate.

(a) Calculate the traditional net present value (NPV) of this investment if it takes place in year 0?
(b) What is the net present value if you wait until year 1 and you choose to invest only if the tariff is high? Should you invest in year 0?
(c) What is the real option value (of waiting) at the beginning of year 0? What type of real option is it?

8. **LCOE**

Consider the following elements:

- Plant cost: 1,000 Euro/kW
- Capacity factor: 20 %
- O&M costs: 0.1 Euro/kWh
- Plant life: 25 years

Calculate the LCOE:

(a) with a discount rate of 0 %.
(b) with a discount rate of 5 %.
(c) with a discount rate of 5 % and an escalation rate of 1 %.

References

Ayres, R., Warr, B.: The Economic Growth Engine: How Energy and Work Drive Material Prosperity. Edward Elgar Publishing Limited, Chelthenham, UK (2010)

Bjørndal, E., Bjørndal, M., Pardalos, P., Rönnqvist, M.: Energy, Natural Resources and Environmental Economics. Springer, Berlin (2010). doi:10.1007/978-3-642-12067-1_1

BP: Statistical Review of World Energy 2013. http://www.bp.com/content/dam/bp/pdf/statistical-review/statistical_review_of_world_energy_2013.pdf

Branker, K., Panthak, M.J.M., Pearce, J.M.: A review of solar photovoltaic levelized cost of electricity. Renew. Sustain. Energy Rev. **15**(9), 4470–4482 (2011). doi:10.1016/j.rser.2011.07.104

Danish Energy Agency: Master data register for wind turbines at end of September 2013. 2013. http://www.ens.dk/node/2233/register-wind-turbines

David, P.A.: Path dependence and the quest for historical economics: one more chorus of the ballad of QWERTY. In: Discussion Papers in Economic and Social History, No. 20, Nov 1997

Dixit, A.K., Pindyck, R.S.: Investment Under Uncertainty. Princeton University Press, Princeton (1994)

Fernandes B., Cunha, J., Ferreira, P.: The use of real options approach in energy sector investments. Renew. Sustain. Energy Rev. **15**(9), 4491–4497 (2011). doi:10.1016/j.rser.2011.07.102

Green, R., Vasilakos, N.: Storing wind for a rainy day: what kind of electricity does denmark export? Energy J. **33**(3), 1–22 (2012). doi:10.5547/01956574.33.3.1

Holttinen, H.: The impact of large scale wind power production on the Nordic electricity system. Ph.D thesis, Helsinki University of Technology (2004)

Hotelling, H.: The economics of exhaustible resources. J. Polit. Econ. **39**(2), 137–175 (1931)

IEA: Harnessing variable renewables. OECD Publishing (2011). doi:10.1787/9789264111394-en

Krohn, S., Morthorst, P.-E., Awerbuch, S.: The Economics of Wind Energy. European Wind Energy Association (2009)

Kula, E.: Economics of Natural Resources and the Environment. Chapman & Hall, London (1992)

OECD/Nuclear Energy Agency: Projected costs of generating electricity (2010). doi:10.1787/9789264084315-en

Oswald, J., Raine, M., Ashraf-Ball, H.: Will british weather provide reliable electricity? Energy Policy **36**, 3212–3225 (2008)

Roth, I.F., Ambs, L.L.: Incorporating externalities into a full cost approach to electric power generation life-cycle costing. Energy **29**(12–15), 2125–2144 (2004). doi:10.1016/j.energy.2004.03.016

Salomon, S., Qin, D., Manning, M., Marquis, M., Averyt, K.B., Tignor, M., Miller, H.L.: Contribution of working groups I to the Fourth assessment report of the Intergovernmental Panel on Climate Change. IPCC (2007)

Slade, M.E.: Trends in natural-resource commodity prices: an analysis of the time domain. J. Environ. Econ. Manage. **9**(2), 122–137 (1982)

Thomas, G.E., Stamnes, K.: Radiative Transfer in the Atmosphere and Ocean. Cambridge University Press, Cambridge (2002)

United States Census Bureau: World population: 1950–2050, Dec 2008. http://www.census.gov/population/international/data/idb/worldpopgraph.php

Willliam, E.D., Purves, K., Sadava, D.: Human Nutrition and Diet: Recommended Intakes of Nutrients. Freeman, New York (2004)

Zacklan, A., Abrell, J., Neumann, A.: Stationarity changes in long-run fossil resource prices. Discussion Paper DIW Berlin, (1152) (2011)

Chapter 2
Fossil Energy Systems

Abstract This chapter gives an overview of the fossil energy sources, which undoubtedly still are dominating the supply of energy. We introduce here how coal, oil and gas are explored, recovered and used for energy production. The most conventional resources and production methods are described, and at the end of the chapter we introduce some of the unconventional fossil energy sources which have a promising resource potential. Fossil fuels are non-renewable resources and therefore represent a limited source of energy. The global reserves will be assessed and the cost of continued production of energy based on these diminishing resources will be evaluated. But first, we start with a historical review of mankind's involvement with fossil deposits and use as an energy source.

Keywords Coal · Oil and gas · Peak oil · Non-conventional fossil resources

2.1 Historical Development

We need energy because of the work it can do for us, and so we have developed elaborated supply chains to obtain fuel cheaply and reliably. Our world is not only served by those supply chains, but also shaped by them. Every time we have switched to a new primary fuel, society has undergone some fundamental reorganization as a result.

For 40,000 years, or more, controlled fire was our primary source of energy. We gathered sticks and twigs and stockpiled *firewood*, and then burned it to convert the energy stored in the wood into heat and light. Then, around 4,000 BC, we discovered how to harness the power of animals, for plowing or driving water pumps. This resulted in a great leap forward, into the agrarian age, and allowed us to convert the chemical energy in *hay*, the food eaten by the ox, into mechanical energy.

Records suggest that *coal* was first introduced as fuel in Scotland in the 9th century by monks to heat their abbeys. Over the next centuries, coal power became adopted

P. A. Narbel et al., *Energy Technologies and Economics*,
DOI: 10.1007/978-3-319-08225-7_2,

by brewers and smiths. Demand soon grew to such an extent that by the 14th century, a coal trade and a new energy supply chain had developed in England. However, an environmental conflict well known to our generation, soon arose. On the one hand, the government encouraged use of coal in breweries and smithies in order to conserve the rapidly diminishing forests needed for ship building. On the other hand, the pollution from coal burning soon became so extensive that the government also put restrictions on this energy source.

By the mid 16th century, the burgeoning iron trade in England increased the demand for coal and soon depleted the easily accessible coal layers near the surface of the ground. So coal miners had to do what oil drillers centuries later would do, go to ever deeper pits. A new challenge then arose, these deep pits needed to be drained of water on a nearly constant basis. Although crude pumps based on horsepower were available, the increasing demand for coal could hardly be met by the advent of the industrial age in the eighteenth century. But a radical new technology emerged that saved the day when the *steam engine* was invented. The steam being produced by burning coal, the steam engine was of course not a new energy source, but rather an energy conversion device. It was soon applied to a wide variety of innovations, and the world switched definitely from wood to coal, its first major fuel substitution, and nothing was ever to be the same again.

Oil has been an important commodity for mankind for thousands of years, but for most of this time oil has been of animal origin. Its most widespread use was in lamps and not as an energy source. Up until 150 years ago, whale oil was the world's primary illuminating fuel. But quality oil extracted from whale blubber was getting extremely expensive in the 1850s as the whales themselves were becoming more and more scarce. Fortunately, the new energy source, fossil oil, was discovered when the demand for lamp oil was at its peak and the supply rapidly falling. At first the new petroleum based oil was distilled from bitumen, a sticky, black and highly viscous liquid derived from tar sands found in Pennsylvania, USA and Ontario, Canada. The new fuel was called kerosene and was cheap enough that nearly everyone could afford it. But as an energy source substantial enough for heating and to run engines, the bitumen distillation was clearly too complicated and limited.

Soon, entrepreneurs and industrialists turned their minds to how they could get out oil from the ground in greater quantities to meet the world's demand. The first, but not very successful well was drilled in Hamilton, Ontario in 1858. It didn't go deep enough to yield much oil. In Titusville, Pennsylvania, prolific flows of subsurface oil was observed at the same time. It was therefore decided to drill into the formation using a derrick originally constructed to bore for salt and using a steam engine to power the drill bit. In 1859 the first underground oil well was a success as it struck oil about 70 feet or so beneath the surface.

Since then millions of oil and gas wells have been drilled globally, most of them now abandoned. But new wells are drilled all the time, and the amount yet to be recovered no one knows. However, the fossil energy resources must be limited, and it is a curious irony that the petroleum reserves which have taken the planet several hundred million years to produce, may be exploited in a couple of centuries, leaving nothing to future generations.

2.2 Coal

Globally, fossil fuels are providing about 80 % of our energy needs: 32 % from oil, 27 % from coal and 21 % from natural gas (in 2010). Coal has for centuries been used for domestic heating, but today, most of the coal is used for fueling electric power plants. In fact 41 % or 8,700 TWh of the electric energy produced globally in 2010 came from coal-fueled plants. Coal is one of the most abundant fossil fuels on Earth, and usually easier and cheaper to explore and produce than other fossil fuels such as oil and natural gas.

Coal formation began during the Carboniferous Period which spanned 360 to 300 million years ago. The build-up of silt and other sediments, together with movements in the Earth's crust—known as tectonic movements—buried wetland forests, swamps and peat bogs, often to great depths. With burial, the plant material was subjected to high temperatures and pressures. The plant matter was protected from biodegradation and oxidation, usually by mud or acidic water. This caused physical and chemical changes in the plant material, transforming it into peat and then into coal in a process called carbonization.

Several types of coal exist. According to their rank or grade:

- *Peat*, considered to be a precursor of coal, has industrial importance as a fuel in some regions.
- *Lignite* or brown coal, is the lowest rank of coal and used almost exclusively as fuel for electric power generation.
- *Sub-bituminous coal*, whose properties range from those of lignite to those of bituminous coal, is used primarily as fuel for steam-electric power generation.
- *Bituminous coal* is a dense sedimentary rock, usually *black*, is used primarily as fuel in steam-electric power generation. It may also be transformed into *coke*.
- "*Steam coal*" is a grade between bituminous coal and anthracite, once widely used as a fuel for steam locomotives.
- *Anthracite*, the highest rank of coal, is a harder, glossy black coal used primarily for residential and commercial space heating.

Coal is composed primarily of carbon along with variable quantities of other elements, chiefly hydrogen, oxygen, nitrogen and small amounts of sulfur. The composition depends on the rank and where it has been mined (see Table 2.1). The *calorific value* Q or energy content of coal is the heat liberated by its complete combustion with oxygen. Q is a complex function of the elemental composition of the coal. Q can be determined experimentally using calorimeters, and is of the order of 30 MJ/kg. Dulong has suggested following empirical formula to determine the energy content of coal:

$$Q(\text{MJ/kg}) = 337\text{C} + 1442(\text{H} - \text{O}/8) + 93\text{S} \tag{2.1}$$

where C is the mass percent of carbon, H is the mass percent of hydrogen, O is the mass percent of oxygen, and S is the mass percent of sulfur in the coal.

Table 2.1 Characteristics of different types of coal (Lindner 2007)

Type	Carbon MJ/kg	Hydrogen MJ/kg	Oxygen MJ/kg	Heat content MJ/kg
Lignite	60–75	6.0–5.8	34–17	<28.5
Flame coal	40–45	6.0–5.8	>9.8	<32.9
Gas flame coal	82–85	5.8–5.6	9.8–7.3	<33.9
Gas coal	85–87.5	5.6–5.0	7.3–4.5	<35.0
Fat coal	87.5–89.5	5.0–4.5	4.5–3.2	<35.4
Forge coal	89.5–90.5	4.5–4.0	3.2–2.8	<35.4
Nonbaking coal	90.5–91.5	4.0–3.75	2.8–3.5	35.4
Anthracite	>91.5	<3.75	<2.5	<35.3

Fig. 2.1 Industrial complex with coal-fired power plant in Ontario, Canada

2.2.1 Coal Plants

In a fossil-fueled power plant, the chemical energy stored in fossil fuels such as coal and oxygen of the air is converted successively into thermal energy, mechanical energy and, finally, electrical energy for continuous use and distribution across a wide geographic area. The electrical power output from a fossil-fueled power plant typical ranges from 500–1,000 MW by burning 250–500 tonnes of fuel per hour under full load.

The second law of thermodynamics states that any closed-loop cycle can only convert a fraction of the heat produced during combustion into mechanical work. The rest of the heat, called waste heat, must be released into a cooler environment during the return portion of the cycle, (see Fig. 1.1). Coal-fueled power plants therefore have

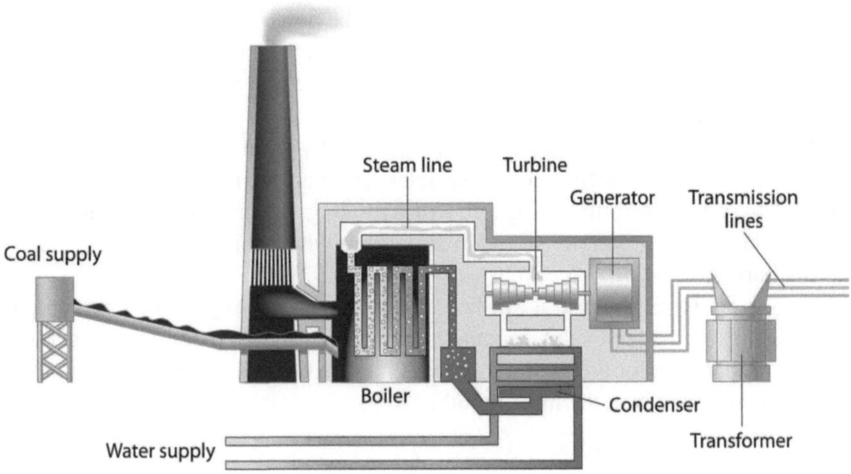

Fig. 2.2 Electricity from coal—flowsheet

two waste problems: the emission of greenhouse gases and waste heat. However, some of the latter may be used for heating homes or for industrial purposes etc.

The total energy production cycle in a coal-fueled power plant is complex and takes place in several stages. The major factory components are:

- Fuel transport, storage and preparation
- Burner
- Boiler
- Steam turbine
- Gas turbine
- Condenser
- Cooling tower
- Generator
- Emission control

Figure 2.2 displays the main components of a typical steam-cycle a coal power plant. Below we will give a short outline of each of the components.

2.2.1.1 Fuel Transport, Storage and Preparation

Coal is delivered from coal mines to power plants by rail, trucks or slurry pipe lines. For coastal or riverine plants it may also be delivered by ship or barge. Some plants may be built near coal mines so that the coal can be delivered by conveyor belts. When coal arrives by train, it is usually carried by a "unit train" with a total length of nearly 2 km, containing one hundred wagons, each wagon carrying 100 (metric) tonnes. The wagons or trucks are emptied by a rotary dump to conveyer belts transporting the coal to a stockpile or delivering it directly to the power plant. Colliers which are cargo ships carrying coal, can hold 4×10^4 tonnes of coal and may take several days to unload. Most plants store fuel for at least 30 days, which may amount to as much as 10^6 tonnes of coal.

The coal lumps delivered from a mine are usually from a few cm to ten cm of size and must be pulverized down to particles less than 1 mm before it is fed to the burner. This is done in a pulverizing mill with a rotating ring or a rotating hammer. The pulverized coal is temporarily stored in silos from which it is blown pneumatically into the burner.

In many countries the coal is washed at the mine, removing its mineral content, thus reducing its ash and sulphur content and improving its heating value per unit mass. In preparation for the washing process, the coal lumps may be crushed to less than 1 cm size already at the mine mouth.

2.2.1.2 Burner

The burner mixes air and the coal powder, and ignites the mixture by a spark-ignited oil jet until the flame is self-sustaining. The ashes produced consist of mineral matter, 90 % is *fly ash* which is drawn out of the burner by natural draft and is later captured in particle collectors. About 10 % of the mineral matter falls to the bottom of the boiler as *bottom ash*. When mixed with water, this forms a wet sludge which is carried away.

Some coal power plants use a cyclone furnace, a kind of combustor that can also effectively burn larger pieces of coal. It is formed as a barrel with water-cooled walls. The advantage of this burner is that the majority of the mineral matter forms a molten ash, called *slag*, which is drained into the bottom of the boiler and removed there, while only a small portion exits the boiler as fly ash. The heat is carried away with the hot combustion gases to the main boiler. The disadvantage of the cyclone burner is the high temperature of the furnace which produces large quantities of nitrogen oxides (NO_x).

2.2.1.3 Boiler

While the burner is specific for the fossil fuel used, the boiler and the other major components in a thermal power station is similar for coal-, oil- or natural gas-fueled plants. The boiler is a central component in all kinds of fossil-fueled power plants which produce steam to run turbines. In a coal-fueled plant, the burner and the boiler is often integrated as shown in Fig. 2.3.

Heat is transferred by thermal radiation from the pulverized coal burner to the boiler *water wall* where riser tubes are circulating water which is vaporized into steam in a *steam drum* sitting on top of the water wall. The steam temperature in the steam drum is typically 370 °C and the size of the steam drum is 30 m in length and 5 m in diameter. The saturated steam is heated further in a *superheater* to 565 °C and

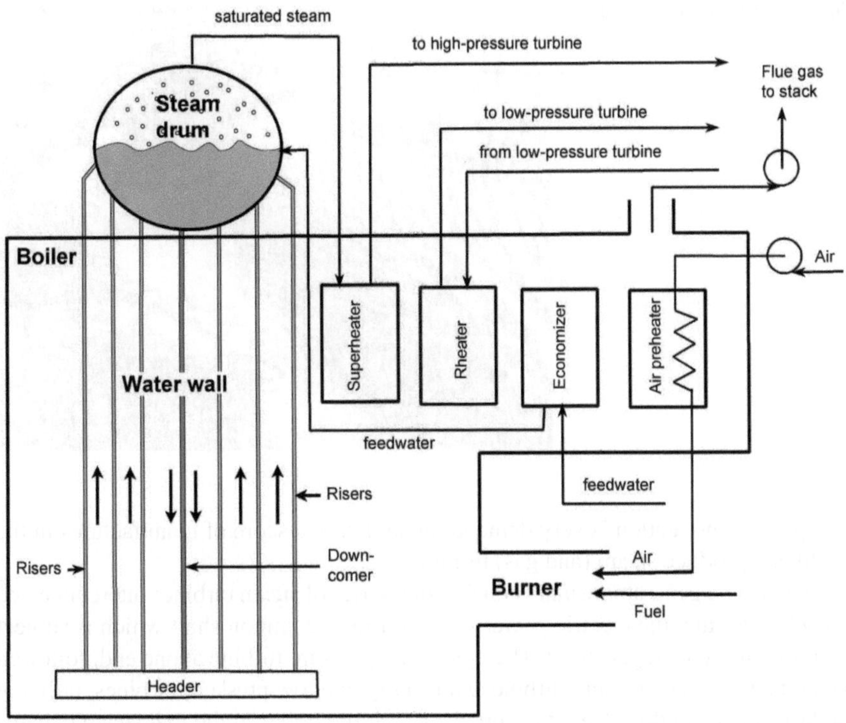

Fig. 2.3 The principle of a combined coal burner and boiler

a pressure of 24 MPa, which is above the critical temperature and pressure for water ($T_c = 374$ °C and $p_c = 22$ MPa). This high pressure steam is then passed on to high-pressure *steam turbines* and the exhaust steam from the turbine is condensed to water in the *condenser* and then passed back to the boiler. The water flow through the boiler is typically 400 l/s for a 500 MW plant at full load. It should also be mentioned that the burner/boiler contains some auxiliary components (air preheater, economizer, reheater) in order to improve the efficiency of the steam generating process.

2.2.1.4 Turbines

A photo of a steam turbine is shown in Fig. 2.4. The purpose of the turbine is to convert the *heat energy to mechanical energy* which can run the *electric power generator* (see Sect. 2.2.1.7).

The steam turbine is the most complex piece of machinery in the power plant. Just as water is pushing the blades of a water wheel in a hydroelectric power plant, a steam jet is pushing the blades of the steam turbine. Considering that the rotating blades of the turbine are subject to very high pressure and temperature, making a

Fig. 2.4 Inspection of a steam turbine

leak proof construction is very demanding, and only a score of manufactures in the world can produce steam (and gas) turbines.

The turbine generator usually consists of a series of steam turbines interconnected to each other and the electric power generator on a common shaft which is subject to enormous centrifugal stress. There is a high pressure turbine at one end, followed by an intermediate pressure turbine and usually two low pressure turbines, utilizing the large pressure drop from the superheater of the boiler to the exit steam exhaust from the total turbine generator. The entire rotating mass may be over 200 tonnes and 30 m long.

There are two different types of steam turbines: the *impulse turbine* and the *reaction turbine*. They differ in the geometry and the angle at which the steam impinges on the blades of the turbine. Both systems have a set of rotors and stators. A schematic drawing of the two turbines is shown in Fig. 2.5. The steam jet is accelerated through a nozzle to a supersonic velocity of about 1,650 m/s and rotates the first rotator. The purpose of the stator is to lead steam on to the next rotor. Each set of rotor and stator is called a stage. When the pressure drop available is large, it cannot all be used in one turbine stage. A single stage utilizing a large pressure drop will have an impractically high peripheral speed of its rotor. This would lead to either a larger diameter or a very high rotational speed. Therefore, machines with large pressure drops employ more than one stage. As steam moves through the system and loses pressure and thermal energy, it expands in volume, requiring increasing diameter and longer blades at each succeeding stage to extract the remaining energy. Turbines can have several dozens stages.

In a *gas turbine plant* where oil or natural gas is used as fuel, the hot combustion gases are directly used to drive a gas turbine, rather than transferring heat to steam and driving a steam turbine. This requires a different turbine, appropriate for the much higher temperatures, 1,100–1,200 °C for the combustion gases, which is the maximum tolerable temperature for the steel alloys used in the turbine blades. The working fluid, composed of nitrogen oxides, water vapor, excess oxygen and carbon

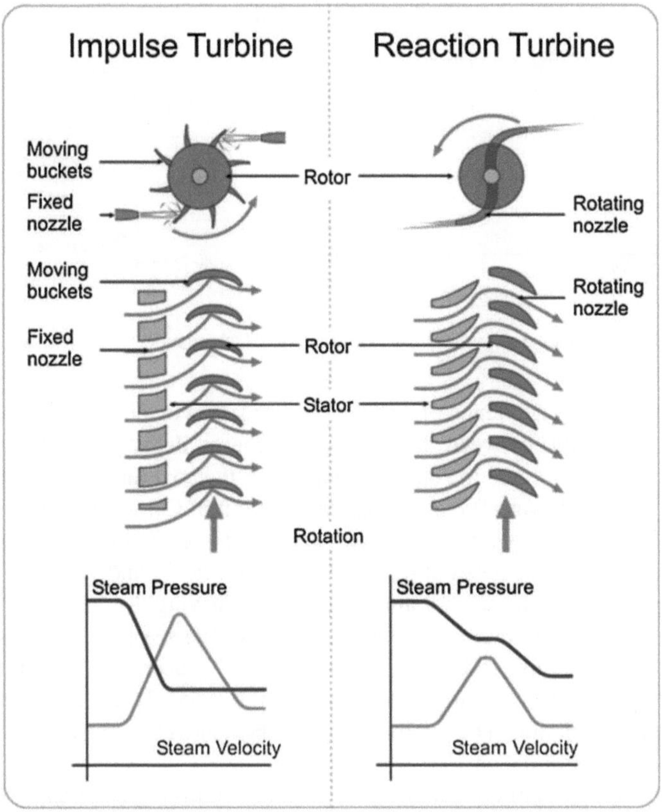

Fig. 2.5 The principle of an impulse turbine to the *left*; reaction turbine to the *right*

dioxides is not returned to the combustion chamber as is the steam in steam turbines, but instead vented into the atmosphere.

2.2.1.5 Condenser

According to the second law of thermodynamics no heat engine cycle can transfer heat from a hot reservoir and transfer it entirely into mechanical work without at the same time delivering heat to a colder reservoir. For a steam turbine, heat is rejected into the environment either in a steam condenser and/or vented into the atmosphere as hot flue gas in the smoke stack. For both cases, the receiving body (the cold reservoir) has a typical temperature of $T_C = 25\,°\text{C} = 298$ K. The hot reservoir which is delivering heat is the superheater of the boiler operating at $T_H = 565\,°\text{C} = 838$ K. The maximum theoretical efficiency of a heat engine operating between these two temperatures is given by the *Carnot cycle* which has an efficiency $\eta = (T_H - T_C)/T_H = 64\ \%$.

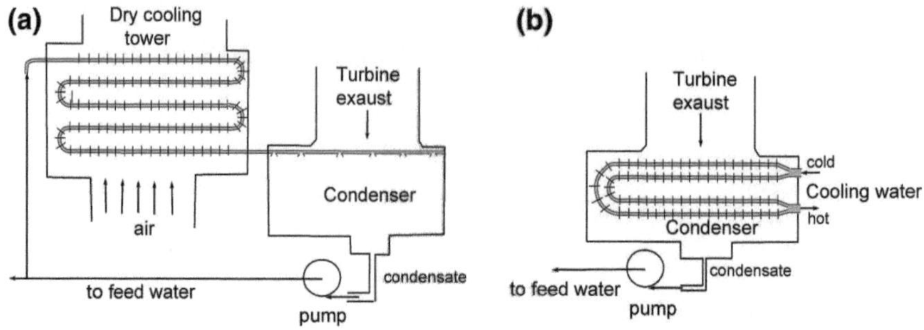

Fig. 2.6 **a** Direct contact condenser. **b** Surface condenser (schematic)

However, the typical efficiency of a steam power plant is in the 33–40 % range. This efficiency is the ratio between the mechanical energy delivered by the turbine to the electrical generator and the chemical energy of the coal fuel fed into the burner. The reason for the lower efficiency than the maximum Carnot efficiency is due to heat being lost through walls and pipes, frictional losses and heat escaping to the atmosphere with the flue gases at temperatures between T_H and T_C.

It is important that the temperature of the water leaving the condenser is as low as possible before it is returned to the steam cycle. We see from the efficiency of the Carnot cycle that the efficiency increases when T_C decreases. When the steam condenses to water, the pressure is also considerably reduced creating a near vacuum that helps the circulation. The cooling agent in the condenser is usually water from a nearby lake, river or ocean.

There are two types of condenser: *direct contact* and *surface contact* condensers. A direct contact condenser is depicted in Fig. 2.6a. The turbine exhaust passes an array of spray nozzles through which cooling water is sprayed, condensing the steam by direct contact. The warm condensate is pumped into a *cooling tower* where updrafting air cools the condensate that flows in the tubes. The cooled condensate is recycled into the spray nozzles. Because the cooling water is in direct contact with the feed water that goes back to the boiler, its purity must be maintained, just like that of the feed water. This purification may be expensive and the majority of power plants therefore use a surface contact condenser as shown in Fig. 2.6b. This is a shell type condenser where the turbine exhaust passes an array of tubes in which the cooling water flows. Very large volumes of steam need to be condensed and the contact surface area may be as large as 50,000 m^3 and the flow rate of the cooling fluid 15 m^3/s for a 500 MW plant at full load.

2.2.1.6 Cooling Tower

Cooling towers are heat removing devices used to transfer rejected heat to the atmosphere. Heat removal by condensers described above where the excess heat is transferred to the surface in rivers or lakes represents a *thermal pollution* of the environment and possible harm to aquatic organisms. Many countries therefore man-

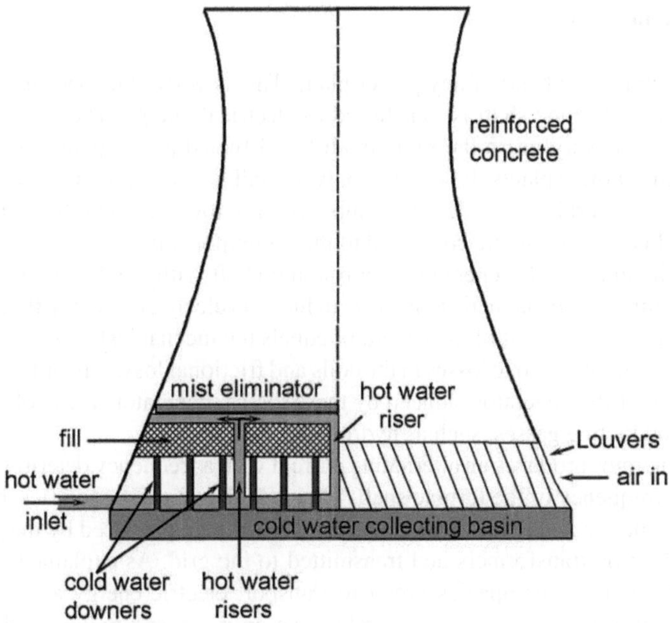

Fig. 2.7 Wet cooling tower, schematic

date that heat rejection occur into the atmosphere via cooling towers. These towers represent a visible landmark of thermal power plants. They have a spool-like structure with hyperbolic vertical cross section which is favorable from a structural standpoint and makes the sometimes 100 m high construction more resistant to strong winds.

There are two types of cooling towers: *wet* and *dry*. A schematic drawing of the wet cooling tower is shown in Fig. 2.7.

Hot water from the condenser is sprayed over a lattice work of closely spaced slats or bars, called *fill* or *packing*. Outside air is drawn by natural draft through louvers surrounding the bottom of the tower. Heat is transferred into the air directly or by evaporating some of the circulating water there by taking the latent heat from the steam flow. This results in a mist plume which is often mistaken as pollution, but it only consists of water droplets or ice crystals in the winter. The amount of water evaporated from a wet cooling tower in a 500 MW plant in a warm climate is of the order 10^7 m^3/year. This must be replaced by surface water from a nearby river or lake, or by municipal water.

In a dry cooling tower, the recirculating water flow through finned tubes over which cold air is drawn and all the heat rejection goes into the air without any evaporation of the circulating water. The dry tower is less expensive to maintain, but the back pressure and the temperature of the returned water may be higher, thereby reducing the efficiency of the plant. In arid areas with little cooling water available, the dry cooling tower must be employed.

2.2.1.7 Generator

The generator is the heart of any power plant. This is where the mechanical energy of the rotating turbine shaft is transferred to electrical energy. The construction of the generator is very much the same in all fossil-fueled power plants as well as in hydroelectric power plants. It occupies only a small area compared to other components (burners, turbines, condenser, pumps, cooling tower etc.) in the plant, and its noise level is also negligible compared to these components.

The principle for all generators is a rotating shaft with conducting coils which rotates within a static magnetic field. This induces an electric current within the coils. The electric power output of the generator equals the mechanical power input of the shaft minus minor resistive losses in the coils and frictional losses. In order to prevent overheating of the generator induced by these losses, generators are cooled by high thermal conducting gasses, such as hydrogen or helium.

The generator produces an alternating current with a frequency determined by the rotational frequency of the turbine shaft. This is 50 Hz in most countries, but 60 Hz in North-America and a few other countries. The voltage produced by the generator is stepped up by transformers and transmitted to the grid. As explained on page 9 in Sect. 1.2.1 utility companies prefer to transport electric energy at high voltage in order to use thinner cables and thereby save on metal and reduce cable weight. Long-distance electrical transfer lines may have a voltage of 400 kV or more. At the user side of the line, the voltage is reduced again by transformers in several steps to 110 or 220 V.

2.2.1.8 Emission Control

The burning of coal in a power plant results in a considerable amount of pollutants that have to be controlled to protect human health and the environment. The emissions, as well as the overall efficiency of the plant, depends on which temperature the boiler/turbine is operated. In a *super critical power plant* this temperature is around 580 °C, compared to 450 °C for a *subcritical plant.* A supercritical plant is much more efficient than a subcritical plant, producing more power from the less coal and with lower emissions.

Most countries have environmental regulations requiring the operator of the power plant to install emission control devices for a number of pollutants escaping the burner to the smoke stack. The most important are:

- Products of incomplete combustion (PIC)
- Carbon monoxide (CO)
- Particulate matter (PM)
- Sulfur (SO_2)
- Nitrogen oxide (NO_x)

The control of PIC and CO is relatively easy to accomplish. If the fuel and air is well mixed, and the fuel is burnt in excess air, the flue gas will contain very little, if any, PIC and CO. It is also in the interest of the power plant to reduce these emissions because a complete burn-out increase the thermal efficiency of the plant. Occasionally, especially during start-ups and component breakdowns, when the flame temperature and the fuel-air mixture is not optimal, PIC emanates through the smoke stack as black visible smoke.

Particles, also called *particulate matter* (PM), would be the most predominant pollutant emanating from the power plant if not controlled at the source. This stems from the fact that coal, and also oil, contains a significant fraction of incombustible *mineral matter*. Depending on the quality of the coal, the mineral matter may amount to typically 10 % by weight. This results in particles in the fly ash, including a host of toxic metals, such as arsenic, selenium, cadmium, manganese, chromium, lead and mercury. In addition there is nonvolatile organic matter (soot), including polycyclic aromatic hydrocarbons. For that reason, most countries have instituted strict regulations on particle emission from power plants. In older plants and plants with cyclone burners, the mineral accumulates at the bottom of the boiler as bottom ash and is discarded as solid waste and sluiced away.

The modern burners using pulverized coal produce as much as 90 % fly ash and these use two kinds of particle collectors. The *Electrostatic Precipitator* works on the principle of charging the particles in a corona discharge and then attracting the charged particles to a grounded plate where the charge is neutralized. The neutral particles are periodically shaken off the plates and collected as solid waste. An alternative particle collector, especially effective for collecting small particles, less than 2.5 μm in diameter, is the *Fabric Filter* or *Baghouse*. It works on the principle of a domestic vacuum cleaner. Long cylindrical tubes (bags) of a special heat resistant fabric are hung up-side down. Each bag has 12–35 cm diameter and 4 m height. The baghouse may contain several thousands tubes. The fly ash is sucked through the tubes, and the particles in the fly ash are stopped by the fabric of the tube walls. The collected particles can be removed by mechanical shaking of the ram which holds the bags.

The coal usually contain 0.7–3 % sulphur by weight. The sulphur stems from cellular sulphur in plants and organisms and is carried on to the fossilized coal deposits. Without sulphur emission control, the sulphur which is oxidized to SO_2 in the burner, will be emitted through the smoke stack into the environment. Emission of SO_2, minor quantities of SO_3 and sulfuric acid H_2SO_4 from fossil power plants are major contributors to acid precipitation. There are basically three approaches to reducing the sulphur emission: before, during or after combustion of the fossil fuel. Washing the coal at the mine mouth is used in many cases to reduce the mineral matter, including various metal sulfides. However, coal washing introduces a pollution problem of the environment. Removing sulfur oxides after combustion is often done in a so called *Scrubber* in which the flue gas is treated with a sorbent,[1] usually sintered limestone ($CaCO_3$) in an aqueous slurry. The reactions taking place between the SO_2 and the $CaCO_3$ is:

[1] A sorbent is a material used to absorb liquids or gases.

Table 2.2 Air pollution from fossil combustion plants in EU 2008: Mass (g) per energy unit (GJ)

Pollutant	Hard coal	Brown coal	Fuel oil	Gas
CO_2 (g/GJ)	94,600	101,000	77,400	56,100
SO_2 (g/GJ)	765	1,361	1,350	0.68
NO_x (g/GJ)	292	183	195	93.3
CO (g/GJ)	89.1	89.1	15.7	14.5
Organic compounds (g/GJ)*	4.92	7.78	3.70	1.58
Particulate matter (g/GJ)	1,203	3,254	16	0.1
Flue gas volume total (m^3/GJ)	360	444	279	272

*Non-methane

$$CaCO_3 + SO_2 + \frac{1}{2}H_2O \rightarrow CaSO_3 \cdot \frac{1}{2}H_2O + CO_2 \tag{2.2}$$

$$CaSO_3 \cdot \frac{1}{2}H_2O + \frac{1}{2}H_2O + \frac{1}{2}O_2 \rightarrow CaSO_4 \cdot 2H_2O \tag{2.3}$$

The sulphur has thus been transferred from a gas (SO_2) to solid calcium sulfite and sulfate which falls to the bottom of the wet scrubber. The hydrated sulfate is gypsum which can be used as by-product of the power plant.

The other major category of gaseous pollutants which emanates from fossil fuel combustion is nitrogen oxides called NO_x. These are pernicious pollutants because they are respiratory tract irritants, and they also contribute to acid precipitation. The NO_x is produced because coal (and oil) contain organic nitrogen in their molecular structure. When burnt, these fuels produce so-called *fuel* NO_x. In addition, all fossil fuel combustion produce *thermal* NO_x. This is a recombination of atmospheric nitrogen and oxygen under conditions of high temperature. The NO_x may be removed after combustion by *selective catalytic reduction* by injecting ammonia (NH_3) into a catalytic reactor. The following reaction takes place:

$$4NO + 4NH_3 + O_2 \rightarrow 4N_2 + 6H_2O \tag{2.4}$$

Thus NO is reduced to elemental nitrogen which is a natural constituent of the atmosphere. Similar reductions take place for other nitrogen oxides. The catalysts used are a mixture titanium and vanadium oxides.

The major effluent from the coal-fueled power plant is of course carbon dioxide (CO_2) which is the product of burning carbon (coal) in an oxygen rich environment. However, CO_2 is not considered a pollutant. On the other hand carbon dioxide is a "green-house gas" which contributes to global warming, see Sect. 1.5.3. Therefore large projects are underway for capturing and storing carbon dioxide from fossil-fueled power plants (CCS), see Sect. 2.5.

Table 2.2 shows the average relative air pollution from 450 electricity-generating large combustion plants within the European Union.

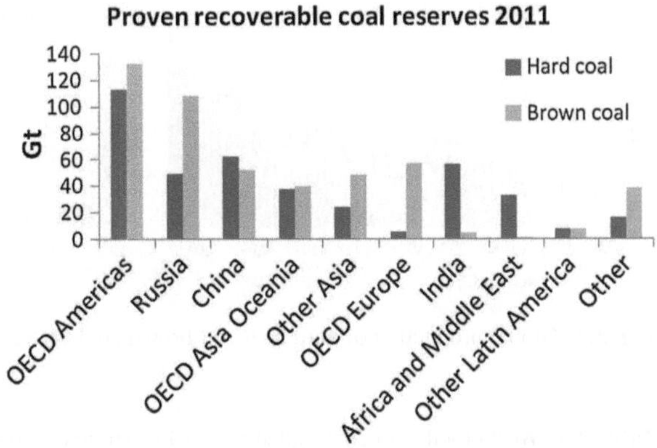

Fig. 2.8 Proven recoverable coal reserves at the end of 2011 (WEC 2013)

2.2.2 Present Use, Resource Considerations and Forecast

As indicated in Sect. 2.1, coal is a major source of energy. At present, almost the totality of the coal used serves as a fuel for electricity and/or heat generation. In few countries, South Africa being one such example, coal is also transformed into a liquid that is further refined into gasoline or diesel fuel.

Coal reserves are large compared to the reserves of other traditional fossil fuels. Figure 2.8 illustrates the state of the proven recoverable coal reserves around the planet, which are considered likely to be recovered under the current market conditions, taking into account the curent mining technologies and their economics. Compared to other fossil fuels, coal is more evenly distributed around the globe and each major region has access to some quantities of coal. Yet, most reserves are located in the OECD countries, Russia, China and India.

At the current rate of extraction, hard coal[2] reserves will last for another 130 years and brown coal[3] reserves for another 270 years. Total coal reserves (including reserves that are not considered economically recoverable at present) are sufficient to last for many more decades.

Overall, the global production and consumption of coal have increased over time (see Fig. 2.9). Paradoxically, the growth rate of coal production exploded since the Kyoto protocol[4] was agreed upon. Almost all of the additional production took place within non-OECD countries, parallel to the growth of their energy needs (these countries did not ratify the Kyoto protocol). The share of the production taking place in OECD countries fell from 56 % in 1971 to 26 % in 2012. More precisely, over two thirds of the growth took place in China as this country has rapidly expanded its

[2] Bituminous, anthracite.

[3] Sub-bituminous, lignite.

[4] This is a global treaty under which many developed countries have accepted to reduce their carbon emissions.

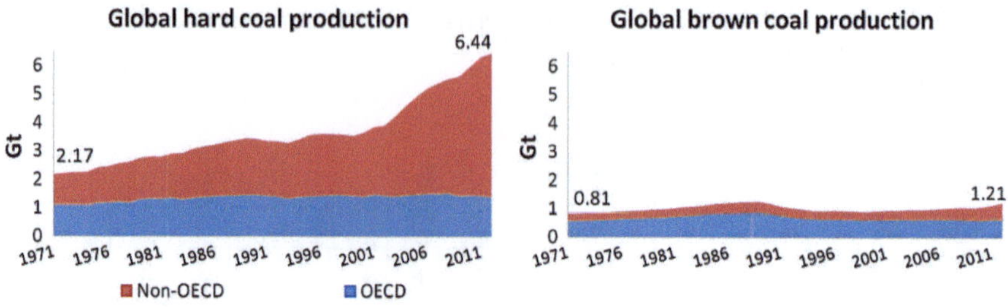

Fig. 2.9 Global production of hard coal and brown coal between 1971 and 2012 (IEA 2012)

Table 2.3 World total hard coal traded in 2010 in Mtonnes (IEA 2011a)

From/to	Japan	Other Asia	OECD Europe	N. America	L. America	Other
Australia	118.6	148.2	18.2	2.8	6.0	2.2
Canada	8.9	13.5	3.6	1.9	2.0	0
Poland	0	0	10.8	0	0	0.1
United States	3	13.4	27.5	9.0	8.7	6.6
PR China	7.0	12.3	0.1	0.1	0	0
Russia	12.1	31.0	46.0	0	0.4	15.9
Indonesia	34.5	291.1	9.1	1.7	0.8	0.9
South Africa	0.3	43.1	17.4	1.3	2.2	3.4
Other	2.2	60.2	49.1	17.1	1.1	15.1

coal industry in the last decades. In 1971, China's share of the world production of coal reached 13 % before steadily climbing to 27 % in 1999 and 42 % in 2010!

The quasi-totality of the extra coal produced or extracted is hard coal. Two reasons can explain that hard coal has expanded more rapidly than brown coal. First, the calorific value of hard coal is higher than the calorific value of brown coal, which generally leads to better economics. Second, the countries contributing most to the growth in coal production have access to large economically recoverable reserves of hard coal.

China is the biggest hard coal producer with 51 % of the global production in 2010. The United States (15 %), India (9 %), Australia (6 %), South Africa (4 %) and Russia (4 %) complete the top six. Germany leads in terms of brown coal production (16 %), followed by Indonesia (16 %) and Russia (7 %) (IEA 2011a).

Consumption generally goes along with production. The countries that have access to large economically recoverable coal reserves are usually the countries that consume most of the coal production. China is the biggest consumer of coal (48 %) followed by the United States (14 %), India (8 %), Russia (3 %) and Japan (3 %) which, together, consume over three quarters of the world coal production (IEA 2012). Large quantities of coal are nonetheless being traded and Table 2.3 illustrates the quantities of hard coal that have been traded between the major actors in 2010.

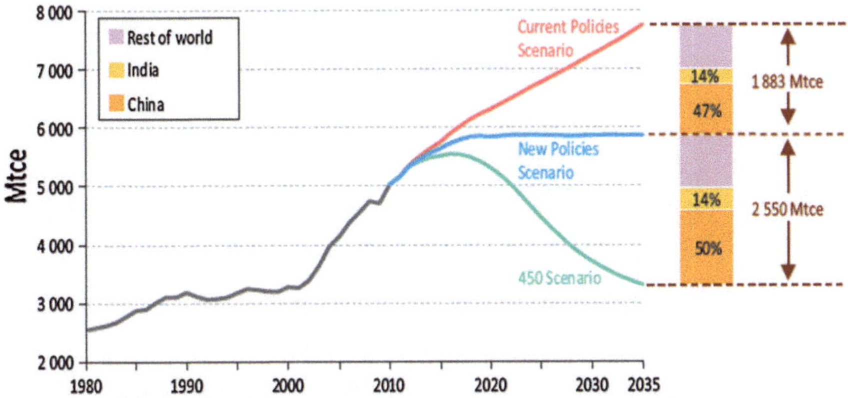

Fig. 2.10 World primary coal demand by region and scenario. *World Energy Outlook 2011*, © OECD/IEA, 2011, Fig. 10.2, p. 358

Over the past decades, environmental pressures have forced the coal power plants to become more efficient in terms of emissions and the average efficiency of new coal based power plant is steadily increasing around the planet. Yet, as these power plants have a long economic life (over 40 years), the average efficiency is only slowly improving.

The future development of coal is highly uncertain. Mostly due to increasing energy needs in developing countries, it is likely that coal will remain a major source of energy in the years to come and that the total installed capacity will expand. The magnitude of development of that energy source will however depend on the scale of future actions implemented to reduce greenhouse gas emissions, because as the 'dirtiest' fuel in terms of emissions, coal is likely to suffer most from climate regulations. Further development of carbon capture and storage technologies at a competitive rate will also influence the use of coal in the future. Figure 2.10 taken from a report by the International Energy Agency perfectly illustrates the uncertainty underlying the use of coal in the future.

The *Current Policies Scenario* assumes that the world follows a path leading to a maximum temperature increase of 2°C, which would require a steep decrease in the use of coal in the next decades. Implementing policies currently in discussion (the *New Policies Scenario*) will lead to a stabilisation in the consumption of coal, whereas a business-as-usual case (the *Current Policies Scenario*) implies that the use of coal will increase significantly in the future.

2.2.3 Cost

The costs presented in this section are based on the data available for black-coal fueled power plants and the emphasis is put on supercritical coal-fueled power plants. This

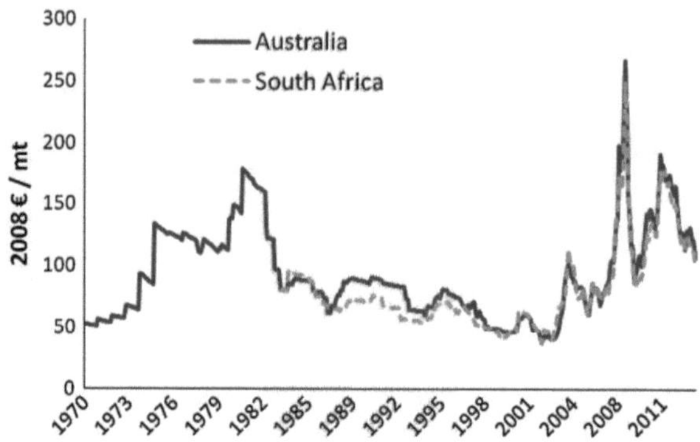

Fig. 2.11 Coal prices in Australia and South Africa in ^{2008}Euro, 1970—July13 (World Bank 2013)

type of power plants operates at higher temperatures and pressures than subcritical power plants, resulting in efficiency improvements and reduced carbon dioxide emissions per tonne of coal used compared to other types of coal-based power plants. Subcritical plants operate at efficiency levels of 39–46 %, which is to be compared to the 30–38 % achieved by subcritical power plants (OECD/Nuclear Energy Agency 2010). With increasing fuel prices and environmental pressures, it is likely that the popularity of the supercritical coal power plants will grow in the future.

2.2.3.1 Resource Cost of Coal

35–70 % of the LCOE for coal-fueled power plants is related to the cost of fuel. The prices of Australian and South African (two major exporters) coal in real terms over the period 1970–2013 are reproduced in Fig. 2.11. It is apparent that the price of coal can change drastically as a response to global events. For instance, the impact of the first oil crisis (1973) can clearly be seen on the graph as the price of coal nearly doubled as a result of a new political will to shift to coal in order to reduce countries' dependency on oil. The expected scarcity of supply meant that prices automatically increased until the supply side could adjust to this increasing demand. The second oil crisis (1979) also led to a surge in coal prices worldwide. More recently, fluctuations in coal prices are due to several distinctive events. First of all, the 2008 energy commodity price surge was also valid for coal. Coal prices later dropped in parallel to the prices of other commodities due to the financial crisis. Second, China started to import large quantities of coal since 2009. Added to an ever increasing coal demand from India since 1990, this increasing demand pressured prices upward. Thirdly, heavy rains and the flooding of some important areas in 2010 have disrupted the coal supply in some parts of Australia, as railway links and open pit coal mines were submerged (Blas 2011), leading to a decrease in the quantities of coal available for export and consequently, to an increase in the price of coal worldwide.

The two time series shown in Fig. 2.11 are highly correlated, showing that disruption in a particular coal market will impact other coal markets as well. In addition, the correlation between coal prices and oil prices is particularly high as well (around 0.76 for the period 1970–2013), indicating that a shock in oil prices will be reflected in the price of coal.

The importance of fuel for a coal-fueled power plant's LCOE combined to large fluctuations in coal prices illustrate one of the risks taken by investors when they invest in such plants. Coal-fueled power plants have an economic plant life of at least 40 years and predictions on future coal prices over the next few decades are needed in order to judge whether an investment is expected to generate positive returns or not. Such predictions are challenging (if not impossible) to make with a decent level of accuracity as many different factors impact the price of coal. For instance, some investors may claim that coal prices will increase in the future as a result of increased demand in China and India. This assumption will be challenged by others which predict that the emergence of unconventional fossil fuels (notably shale gas) will push some countries, e.g. the US, to shift from coal to natural gas, thus reducing global demand for coal and consequently lead to a decrease in the long-term coal prices.

At the moment, coal is one of the cheapest electricity generating technologies and even a moderate increase in the price of coal should not impact this statement. Nevertheless, a last factor, which is also the factor which embodies most uncertainty, is the possible emergence of stringent environmental regulations. Such regulations have the potential to make coal-fueled power plants economically unattractive, because carbon emissions related to coal are particularly abundant (twice those of natural gas for the same quantity of power generated). No one knows today if a global or regional climate agreement will be ratified, and especially its scale, and this uncertainty is perhaps the single most important source of uncertainty for investors. To date, this uncertainty has effectively discouraged many to invest in coal-fueled power plants in areas of the world (e.g. Europe, US) where such regulations may be implemented in the next decade or so.

2.2.3.2 Basic Cost of Coal Power Plants

Capital costs of coal-fueled power plants are low compared to the capital costs of other technologies, which partly explains why coal power is so popular in developing countries. The capital costs of a black-coal fueled power plant ranges between 400–1,800 Euro/kW of installed capacity. Compared to other technologies, the construction time of a coal power plant is relatively long as it takes nearly 4 years to build such a plant (OECD/Nuclear Energy Agency 2010).

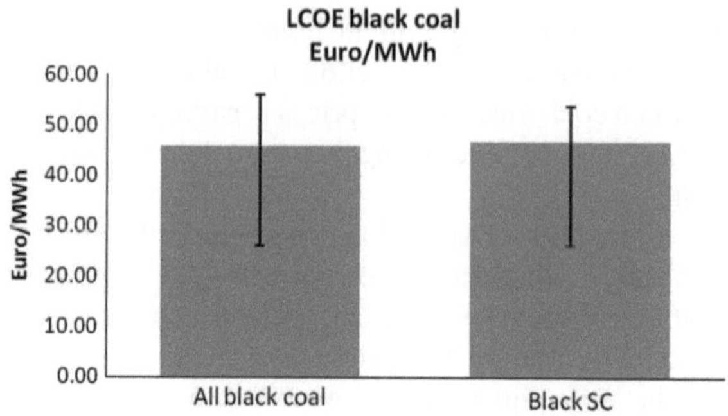

Fig. 2.12 Median levelized cost of electricity for black coal-fired power plants and supercritical coal-fired power plants. The 'error bars' indicate the range of cost

Table 2.4 Levelized cost of electricity for black coal-fueled power plants

	All black coal	Black coal SC*
LCOE (Euro/MWh)	24–53	24–51
Overnight cost (%)	23–53	23–46
O&M costs (%)	4–18	4–16
Fuel costs (%)	34–72	39–72

*Supercritical

2.2.3.3 Electricity Generation Cost of Coal Power Plants

Power from coal is one of the cheapest, if not the cheapest, source of energy in most parts of the world. Low learning rates[5] do not let assume that the levelized costs of coal power will decrease significantly in the future. As long as no price is put on CO_2 emissions,[6] coal will remain highly competitive in the foreseeable future. As mentioned earlier, the big share of the levelized cost is due to the cost of fuel since they account for between 34 and 72 % of the levelized costs of electricity from coal. Capital costs account for a third to about half of the levelized costs of coal power, whereas O&M costs are less important and sum up to less than 10 % of the LCOE. These numbers are summarized in Table 2.4 and Fig. 2.12.

Finally, increasing efficiency can lead to better economics. One such option is to combine electricity and heat (Combined Heat and Power, or CHP) in order to reduce the energy loss in the process of producing electricity.

[5] See Sect. 5.1.4 for a description of learning rate.

[6] See Sect. 5.2.4 for a description of the impact of a carbon price on the cost of coal.

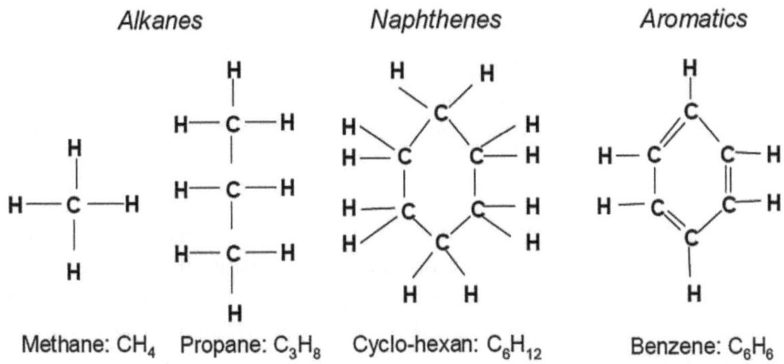

Fig. 2.13 Examples of hydrocarbons in petroleum. Methane and propane are gases under normal conditions (atmospheric conditions). Cyclo-hexan and benzene are liquids

2.3 Oil and Gas

2.3.1 Basic Properties

In this chapter we will generally treat oil and gas together because they have many overlapping properties, and because the exploration for oil and gas usually is one and the same venture.

Most substances normally exist in three *phases*: the solid phase, the liquid phase and the gas phase. Temperature and pressure will decide in which phase a substance exists. Solids have a tendency to keep their shape and volume when subject to moderate external forces. Liquids on the other hand, will change their shape but retain their volume, and gas will change both shape and volume. Nevertheless, there is no clear separation between the three phases, especially not between liquid and gas, which is often denoted by the common name: *fluids*. Both oil and gas are therefore fluids. At the same time they belong to a chemical group called *hydrocarbons* because their molecules mostly consist of hydrogen and carbon atoms, but may also include sulphur, nitrogen, oxygen and metallic compounds. We use the term *petroleum* about the mixture of hydrocarbons that we want to recover from a petroleum reservoir. Depending on temperature and pressure, it may be a liquid, a gas or even a solid phase.

The hydrocarbons found in a petroleum reservoir vary from the simple molecule *methane* (CH_4) with a molecule weight of 16, to naphthenes and polycyclic molecules with a molecule weight of more than a thousand. All molecules have been created through thermal or bacterial decomposition of organic matter subject to high pressure and temperature during millions of years. Figure 2.13 displays examples of molecules which may be found among the hydrocarbons in a petroleum reservoir.

It is often more practical to refer to a special hydrocarbon by the number of carbon atoms in the molecule instead of the exact chemical formula. Methane is then denoted C_1, propane C_3 etc. The components which are liquids under atmospheric or *standard*

Table 2.5 Crude oil classification based on its density

	°*API*	ρ_0 (g/cm^3)
Light oil	>31.1	<0.870
Medium oil	31.1–22.3	0.870–0.920
Heavy oil	22.3–10	0.920–1.000

conditions (1 atm =1.013 $\times 10^5$ Pa and $T = 15$ °C) are described as *crude oil* in commercial contexts. Crude oil may consist of thousands of different molecule types and can vary from a light-brown liquid to a very high viscosity tar-like fluid. Crude oils are often broadly categorized by using properties which are easy to measure in the field, such as the gravity of the oil. For historical reasons the gravity is measured in *degrees API* (American Petroleum Institute) which is defined:

$$^\circ API = \frac{141.15}{\rho_0} - 131.5$$

where ρ_0 is the density in g/cm^3. Crude oil is then classified according to its density as in Table 2.5.

Because of the high temperature and pressure in a petroleum reservoir, the oil usually contains dissolved gas. When the oil is brought to the surface some of the gas will be released, which results in a shrinkage of the oil volume. The gas/oil ratio (GOR) is a dimensionless number equal to the ratio of the released gas volume to the oil volume at the surface (at standard conditions). The GOR and the API is a first indication of the quality of the oil.

2.3.2 Hydrocarbon Accumulations

Several conditions must be satisfied for a hydrocarbon accumulation to be established as a petroleum reservoir. The first is a *sedimentary basin* where a suitable sequence of rocks has accumulated over geological time. Within this sequence there must be a high content of organic matter, *the source rock.* The source rock reaches *maturation* through elevated pressure and temperature, a condition at which hydrocarbons are expelled through a process called *migration* and transferred into a porous type of sediment, the *reservoir rock.* Only if the reservoir is deformed in a favorable shape or if it is laterally grading into an impermeable formation, does a *trap* for the migrating hydrocarbons exist. These conditions are illustrated in Fig. 2.14 and we will describe them in more detail below.

2.3.2.1 Source Rock

In petroleum geology, source rock refers to a rock where hydrocarbons are being generated from organic matter. About 90 % of all organic matter found in sediments is contained in *shale,* which is a fine-grained clastic sedimentary rock composed

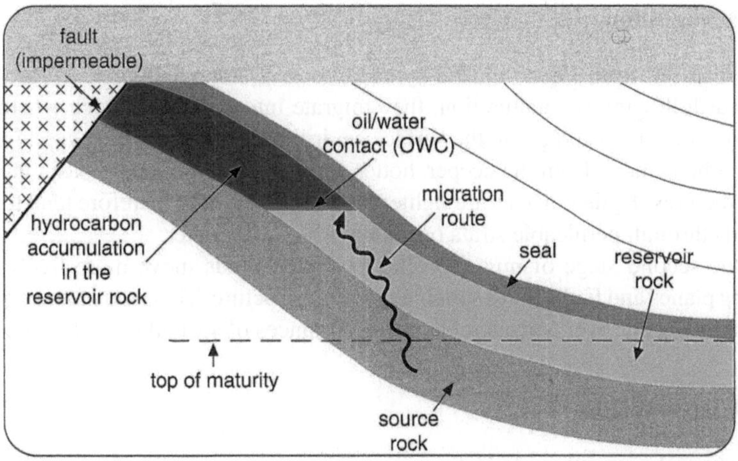

Fig. 2.14 Generation of hydrocarbons in a source rock, migration and trapping in the reservoir rock

of mud that is a mix of flakes of clay minerals. Continuous sedimentation over a long period of time causes burial of organic matter. These organic deposits can be divided into three groups depending on its origin and the type of hydrocarbon it produces. It may originate from *algal remains* deposited under anoxic (lack of oxygen) conditions in deep lakes or ocean. They tend to generate waxy crude oils when submitted to thermal stress during deep burial. An other source rock may be formed from *marine planktonic* and bacterial remains preserved under anoxic conditions in marine environments. Source rocks which are formed from terrestrial *plant material* that has been decomposed by bacteria and fungi under oxic or sub-oxic conditions. They tend to generate mostly gas and light oils when thermally cracked during deep burials. Most coals and coaly shales are this kind of source rock.

2.3.2.2 Maturation

The solid, insoluble organic matter that occurs in source rocks is called *kerogen.* Typical organic constituents of kerogen are algae and woody plant material, as described in the previous subsection. With increasing burial and pressure from later sediments, and increase in temperature, the kerogen within the source rock begins to break down. This thermal degradation or cracking releases shorter chain hydrocarbons from the original large and complex molecules found in kerogen. As a result lighter hydrocarbon molecules are formed. This conversion of sedimentary organic matter in the source rock into petroleum is termed *maturation*, and the important factor in this process is heat. Depending on the temperature in the source rock, the crude oil itself will begin to crack and gas will start to be produced. Initially the composition of the gas will show high content of C4–C10 components (wet gas or condensate), but with further increase in temperature the mixture will tend towards the light hydrocarbons C1–C3 (dry gas).

2.3.2.3 Migration

As we can see from Fig. 2.14, the hydrocarbons do not accumulate in the source rock, but following the maturation, they migrate into a reservoir rock where from we may produce oil and gas to the surface by drilling. During the *primary migration* hydrocarbons move from the deeper, hotter parts of the sedimentary basin into suitable structures. Hydrocarbons are lighter than water and will therefore tend to move upwards through permeable strata originally filled with water.

In the second stage of migration the generated fluids move more freely along bedding planes and faults into a suitable reservoir structure. This *secondary migration* process can occur over considerable lateral distances of several tens of kilometers.

2.3.2.4 Reservoir Rock

Reservoir rocks are either of *clastic* or *carbonate* composition. Both are sedimentary rocks. The former are composed of silicates, usually sandstone, the latter of biogenetically derived detritus, such as coral or shell fragments. There are some important differences between the two rock types which affect the quality of the reservoir and its interaction with fluids which flow through them.

The main components of sandstone reservoirs (siliciclastic reservoirs) is quartz (SiO_2). Chemically it is a fairly stable mineral which is not easily altered by changes in pressure, temperature or acidity of pore fluids. Sandstone reservoirs form after the sand grains have been transported over long distances e.g. by rivers or wind, and have been deposited in particular environments of deposition, such as river deltas, shallow marine sand banks or sheet-like sand bodies from storms or transgression (a rising of the sea level relative to the land).

Carbonate reservoir rock is usually found at the place of formation (in situ). Carbonate rocks are susceptible to alteration by the process of *diagenesis* which are a chemical and physical processes affecting a sediment after deposition.

The pores between the rock components, for example the sand grains in a sandstone reservoir, will initially be filled by *pore water*. The migrating hydrocarbons will displace the water and thus gradually fill the reservoir. For a reservoir to be effective, the pores need to be in communication to allow secondary migration, and also need to allow flow towards the borehole once a well is drilled into the structure. The pore space is referred to as *porosity* in oil field terms. *Permeability* measures the ability of a rock to allow fluid flow through its pore system. Porosity is an important parameter for determining the amount of hydrocarbons stored in a reservoir, and the permeability indicates how easy these hydrocarbons are to produce. A reservoir which has a good porosity but low permeability is termed "tight".

2.3.3 Exploration

Finding a petroleum reservoir may at first seem to be like finding a needle in a haystack. Given the cost of exploration ventures, it is clear that much effort will be expended to avoid failures. A variety of disciplines are involved, such as geology, geophysics, mathematics, geochemistry, to analyze a prospective area. However, on average, even in very mature areas where exploration has been ongoing for years, only every third exploration well will encounter substantial amounts of hydrocarbons. In basins that have not been drilled before the rate of success may be as low as every tenth well. The first indication that an area is a potential candidate for closer geophysical exploration, is the general knowledge the geologists have of the area. Mapping of gravity anomalies and magnetic anomalies may be the first indications of a sedimentary basin. Next, *seismic surveying* is carried out, starting with a coarse two-dimensional (2D) seismic grid, covering a wide area. This is performed to find structures that may be candidates for potential hydrocarbon accumulations. These seismic surveying methods, which are described in more detail below, are based on sending *acoustic pulses* into the strata and recording the reflections. Recently, also *electromagnetic techniques* using electromagnetic pulses have been used.

Acoustic seismic surveys involve generating *sound waves* which propagate through the Earth's rock down to reservoir targets. For land surveys this may be truck-mounted *vibrating sources* or small *dynamite charges* detonated in a shallow hole. The most common marine sources are pneumatic air guns or water guns that expel air or water into the surrounding water column to create an acoustic pulse. The waves are reflected from different strata in the underground and registered on the surface in receivers called *hydrophones*. The reflection data is recorded and stored for processing. The result is an acoustic image of the subsurface which is interpreted by geophysicists and geologists. It is quite clear that this involves a complicated analysis that need highly trained and experienced specialists. Their work is decisive for where the first wells will be drilled.

Seismic surveying has progressed to become one of the most effective methods for optimizing field production.

It is used in:

- *exploration* for delineating structural and stratigraphic traps
- deciding where the first wells will be drilled
- *field appraisal* and *development* for estimating reserves, and drawing up field development plans
- *production* for reservoir surveillance such as observing the movement of reservoir fluids in response to production

Seismic acquisition techniques vary depending on the environment (onshore or offshore) and the purpose of the survey. In an exploration phase a seismic survey consist of a loose grid of 2D lines. In contrast, in an underground appraisal, a 3D seismic survey will be shot. The 3D grid is more closely spaced than the 2D and have both vertical and horizontal positions. In some mature fields a permanent 3D acquisition network might be installed on the seabed for regular (6–12 months) surveillance of the reservoir's response to production. A collection of three-dimensional (3D) seismic data acquired at different times over the same area is called a four-dimensional (4D) seismic acquisition.

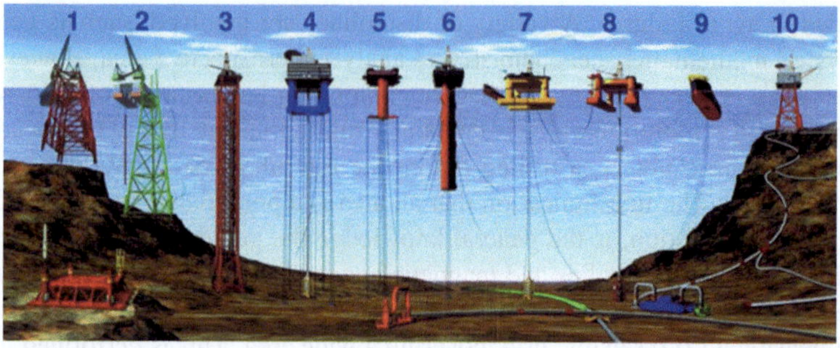

Fig. 2.15 Offshore installations: *1*, *2* Conventional fixed platforms; *3* Compliant tower; *4*, *5* Vertically moored tension leg and mini-tension leg platform; *6* Spar; *7*, *8* Semi-submersibles ; *9* Floating production, storage, and offloading facility; *10* Sub-sea completion and tie-back to host facility (NOAA)

2.3.4 Drilling

The oil rig is what most people first of all associates with the oil and gas industry. The rigs come in many sizes and constructions depending on where the reservoir is located:

- onshore (Saudi Arabia, Texas, Canada)
- shallow water (parts of the North Sea, Gulf of Mexico, Nigeria, Venezuela)
- deep offshore waters (Brazil, the Norwegian Sea, the Barent Sea, Gulf of Mexico) with 500–3,000 m water depths

An artistic review of different offshore installations, both for drilling and production is shown in Fig. 2.15. For offshore drilling rigs, we distinguish between two types: moveable and permanent rigs. The drilling of a well involves a substantial investment from a few million Euro for an onshore well to 100 million Euro plus for a deepwater offshore well. The purpose of the drilling is to collect information about geological formations, find and produce hydrocarbons or to inject water or gas to into the reservoir in order to maintain or increase the reservoir pressure. We may therefore divide the wells into three major classes:

- Exploration wells
- Production wells
- Injection wells

2.3.4.1 The Main Components of the Drill Rig

The principle and the main components of an onshore rig is shown in Fig. 2.16. The basic drilling system is the same for both onshore and offshore rigs and the central unit is the drill tower, called a *derrick*, with a rotary system which rotates the *drill*

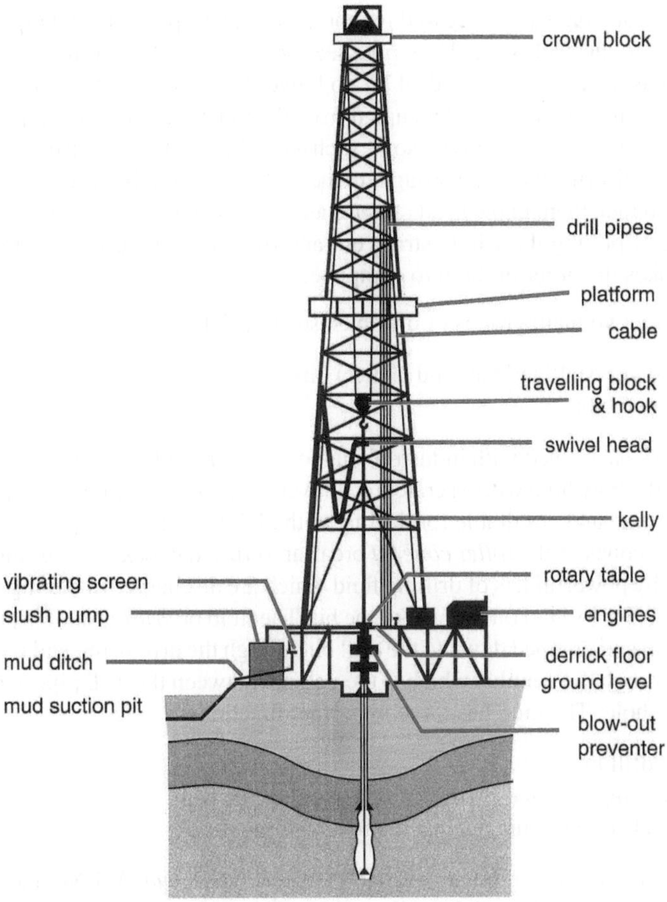

Fig. 2.16 Main components of an oil rig

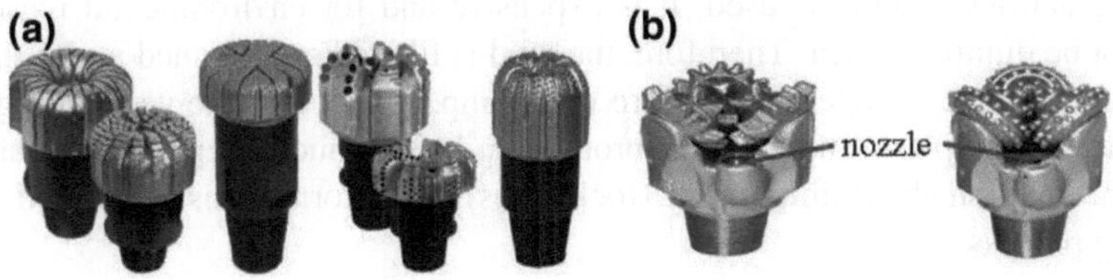

Fig. 2.17 **a** PDC-bits and **b** roller cone bits

string with the *drill bit* at the bottom of the well. The rotation is either provided by a *rotary table* at the derrick floor or for newer and more advanced rigs, by a *top drive* system which is coupled to the top of the drill string and guided by rails so it can move up and down inside the derrick.

After the drilling has progressed for some time a new piece of drill pipe will have to be added to the drill string. Each pipe section is usually 9 m (30 feet) long and the drill string is raised so that the drill bit no longer touches the bottom of the well. In this position the drill string is decoupled from the top drive and a new pipe section is added. Sometimes, for various reasons, such as to change the bit or drilling assembly, the whole drill string has to be brought to the surface. Normally 27 m sections, called a *stand*, and rack them in the mast rather than disconnecting all the 9 m sections. The procedure of pulling the whole string out and running it in again is called a *round trip* and takes the order of 24 h to complete.

There are two main groups of drill bits, see Fig. 2.17:

- PDC (Polycrystalline Diamond Cutter) bits
- Roller cone bits

The *PDS-bit* is lined with industrial diamond cutters and is cutting the rock at the bottom of the bore hole with a scratching movement. It has a high rate of penetration, a long life time and is suitable for drilling with high revolution frequency. The teeth on the three cones on the *roller cone bit* break and crush the rock. The cutting action is supported by powerful jets of drilling fluid which are discharged under high pressure through nozzles located on the side of the bit. The fluid or *drilling mud* is prepared at the platform and pumped down to the drill bit through the drill string and is circulated back up through the annulus which is the volume between the drill pipe and the wall of the bore hole. The mud has three important functions:

- cool the drill bit
- lift the cuttings created by the bit action out of the well
- provide a back-pressure against the reservoir pressure

Two main groups of mud are available; *water-based mud* (WBM) and *oil-based mud* (OMB). The WBM is the most common, and more environmentally acceptable than OMB, which is based on diesel or mineral oil. The correct gravity of the mud is achieved by adding a heavy mineral e.g. baryte ($BaSO_4$). During drilling operations large quantities of mud is used. It is expensive and for environmental reasons it can not be dumped at sea. Therefore, the mud is filtered and cleaned at the surface and kept in large storage tanks before it is pumped back into the well through the drill string. Samples of the cuttings brought up by the mud is kept for analysis and give information about lithography (rock types) of the formations penetrated in the drilling process.

2.3.4.2 Casing and Well Completion

To prevent the bore hole from collapsing the wellbore is lined with *casings* between the drill string and the borehole wall. The casing design resembles a periscope with the largest diameter at the top, and a decreasing diameter downwards towards the top of the reservoir, and cemented to the borehole wall. *Well completion* is the procedure for preparing the well for production. The last section of the casing is the *liner* which perforated so that the reservoir fluids can enter into the production line inside the casings.

A *blowout preventer* (BOP) is sitting between the well head and the platform. The BOP is a series of powerful sealing elements designed to close off the annular space between the pipe and the hole through which the mud normally returns to the surface. As mentioned above one of the purposes of the drilling mud is to provide a hydrostatic head of fluid to counterbalance the pore pressure of the fluids in the reservoir. However, for a variety of reasons the well may "kick", that is formation fluids may enter the wellbore, upsetting the balance of the system, pushing mud out of the hole, and exposing the upper part of the well and the equipment to the higher pressure from the deep subsurface. If left uncontrolled, this can lead to a blowout, a situation where formation fluids flow to the surface in an uncontrolled manner.

2.3.4.3 Directional Drilling and Geosteering

Almost no wells are vertical. Advanced *directional drilling* techniques using rotary steerable systems have been developed. Instead of rotating the whole drill string, mud motors sitting just behind the bit, facilitates the rotation. The motor is run by the circulating mud so that the rotation of the drill string is restricted to the motor section and the drill bit while the rest of the drill string moves with a sliding motion or only rotates with a much lower speed. The state of the art today is that by directional drilling to almost any target in the subsurface, in fact it is also possible to drill upwards. In this way even small pockets of hydrocarbons can be reached.

Directional drilling, often called horizontal drilling, with *geosteering* is probably the one factor that has contributed most to the increased recovery factor over the last two decades. It is based on downhole geological logging, called *measurement while drilling* (MWD). A sonde is placed on the drill string close to the bit. It contains logging tools which measures petrophysical data (porosity, density, resistivity etc.) while drilling. The resistivity measurements enable the driller to steer the bit above the oil water contact (OWC), a technique which is important for producing thin oil zones, see Fig. 2.18. Horizontal drilling and geosteering have today been refined to a technique where it is possible to steer the bit to a target 10 km from the platform with an accuracy of 1 m.

2.3.5 Production

Seismic data acquisition and geological models give important information about the size and the commercial capacity of a reservoir. However, it does not inform us about the flow properties of the reservoir fluids and their interaction with the rock. These fluids (water, oil, gas) are contained under high pressure in a porous network

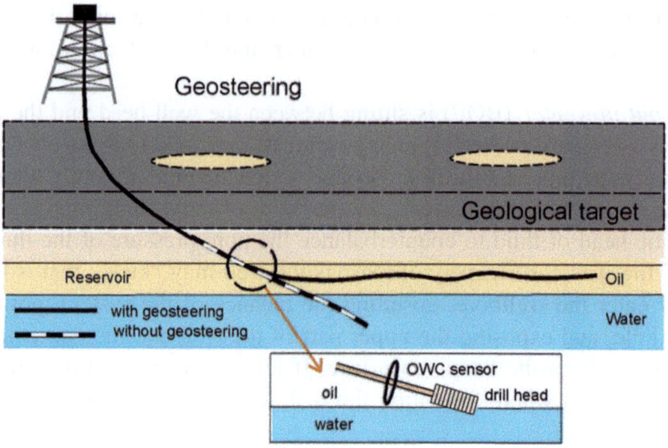

Fig. 2.18 The principle of geosteering. The drill path can be adjusted by the MWD sonde to stay on *top* of the pay zone

with pore sizes ranging from a few μm (10^{-6} m) to several hundred μm. Once the reservoir is opened to the surface through a production well, the high pressure in the reservoir will drive the hydrocarbon towards the well and produce it to the surface. Utilizing the natural pressure of the reservoir called *pressure depletion*, is the simplest of all production mechanisms. The following expansion of the reservoir fluids act as a source of drive energy which support the *primary production* from the reservoir. Primary production means using the natural energy stored in the reservoir as a drive mechanism for production. *Secondary production* implies adding energy to the reservoir by injecting fluids to help supporting the reservoir pressure as production takes place. One also speaks of *tertiary recovery* or *enhanced oil recovery* (EOR *)* which implies adding energy through thermal methods, chemical flooding or injection of gas, for example CO_2 or nitrogen. We will discuss drive mechanisms in primary recovery in somewhat more detail in the next section, and return to secondary recovery (water flooding) later and to tertiary recovery (EOR) in Sect. 2.4.

2.3.5.1 Driving Mechanisms and the Material Balance Principle

A *drive mechanism* can be defined as the energy already present in the reservoir, or which is injected into the reservoir, so that hydrocarbons are produced to the surface. It may be the original reservoir pressure, an expanding gas cap, *natural water drive* when water from an underlying *aquifer* is able to flow into the reservoir, or energy added through injection of water or gas from the surface (secondary recovery). The drive mechanisms may vary from field to field and several drive mechanism may act simultaneously. The *material balance principle* (MBE) states the logical principle that:

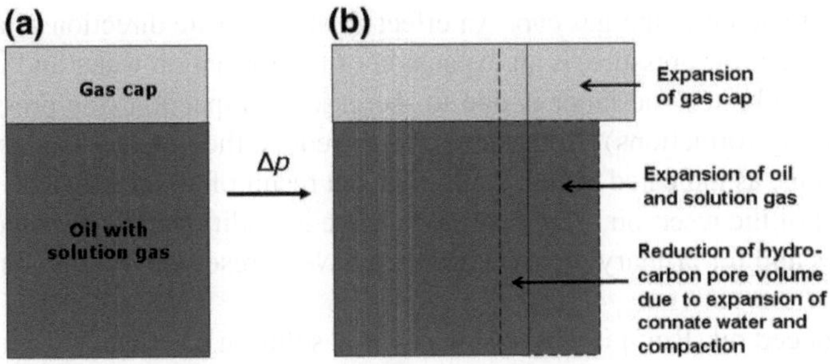

Fig. 2.19 **a** Oil and gas volumes before a production. **b** The volumes after a production which results in a pressure drop

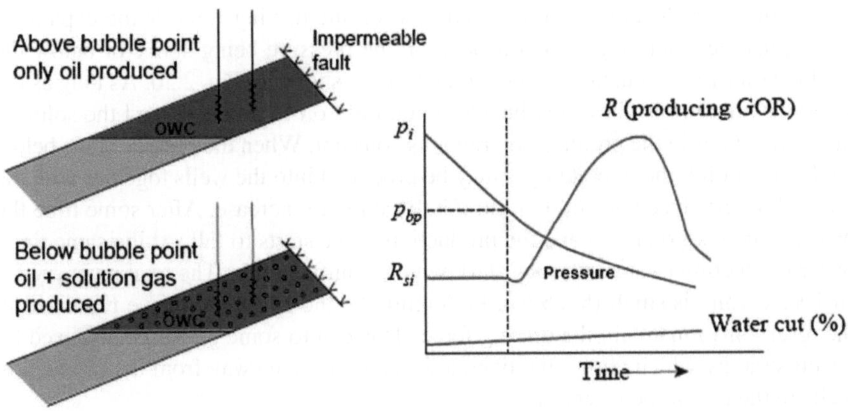

Fig. 2.20 Production profile for solution gas drive reservoir

The mass M_i of the hydrocarbons originally present in the reservoir is equal to the sum of the mass ΔM produced and the mass M remaining in the reservoir.

The MBE-principle may be stated mathematically, which will not be done here, but the equation does not imply any detailed information about the processes going on in the reservoir or any geological models. However, this simple principle has proved to be valuable for:

- forward extrapolation of production curves for oil, gas and water
- identification of drive mechanisms
- history match

Figure 2.19 illustrates a production without any injection from the surface or inflow from an aquifer.

The reservoir contains oil with dissolved gas and a gas cap above the oil, Fig. 2.19a. Oil is produced from the oil zone and there is a pressure drop Δp in the reservoir. As a result of the reduced pressure, the gas cap will expand and so will the oil. If the pressure drops below the bubble point of the oil, some of the solution gas may be liberated and enter the gas cap. An effect in the opposite direction as a result of a reduced reservoir pressure, is an expansion of the formation water and a decrease of the pore volume. The latter is due to increased compaction (the pressure from the overlaying formations). Both these effects reduce the volume available for the hydrocarbons, as indicated in Fig. 2.19b. The net result of all the effects is that oil is pushed out of the reservoir. The material balance is traditionally expressed in terms of volumes and for primary recovery we then have in reservoir volume units:

$$\begin{aligned}\text{Produced oil (Rm}^3) &= \text{Expansion of oil} + \text{solution gas (Rm}^3) \\ &\quad + \text{Expansion of gas cap (Rm}^3) \\ &\quad - \text{Reduction of hydrocarbon pore volume (Rm}^3)\end{aligned} \tag{2.5}$$

The primary production from a reservoir when the driving force is the expansion of oil plus the solution gas, will depend on the pressure being above or below the bubble point p_b. A schematic production curve is shown in Fig. 2.20. As long as the reservoir pressure is above the bubble point, only oil is produced and the solution factor R_s of gas in the produced oil remains constant. When the pressure falls below the bubble point, the liberated gas may be produced into the wells together with the oil, and the produced gas to oil ratio (GOR) starts to increase. After some time the driving force is exhausted and the production curve starts to fall. At the same time, water production (water cut) may start, which is unfavorable. The total recovery for such reservoirs is small (5–25 %). Preferentially, the gas should have remained in the reservoir to maintain the driving force. This can to some extent be achieved by a well strategy which allows the liberated gas to migrate away from the production wells to the top of the reservoir.

2.3.5.2 Waterflood

Waterflood or water displacement is the secondary recovery method in which water is injected into the reservoir formation to displace oil. The water from injection wells physically sweeps the displaced oil to adjacent production wells, see figure 2.21. Zone I is the swept region of the reservoir leaving only unproduceable residual oil, zone II is the water front where both water and oil is flowing, and zone III is the yet unproduced reservoir where the oil saturation is $S_o = 1 - S_c$, where Sc (connate water) is the original water saturation before production starts.

The *recovery factor* E_R is the ratio between the oil produced and the oil originally in place (STOOIP[7]). It depends on three factors:

$$E_R = E_D \times E_A \times E_V \tag{2.6}$$

[7] Stock Tank Oil Originall In Place is a term which normalises volumes of oil contained under high pressure and temperature under ground to surface conditions, 1 bar and 15 °C

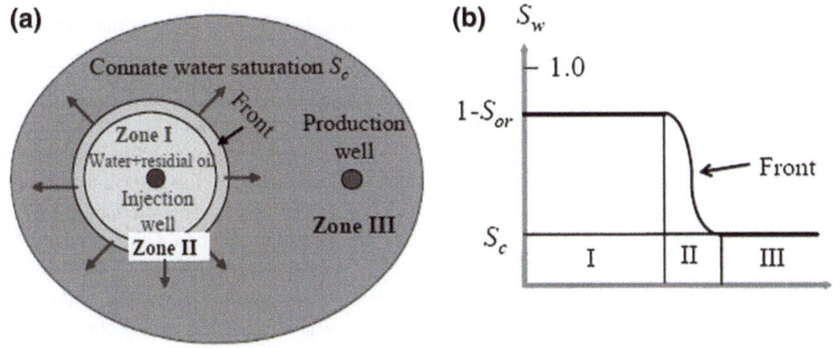

Fig. 2.21 **a** A simple model of production by a waterflood. Water is injected in an injection well and oil produced in the production well. **b** The saturation front shows how the oil is displaced towards the production well

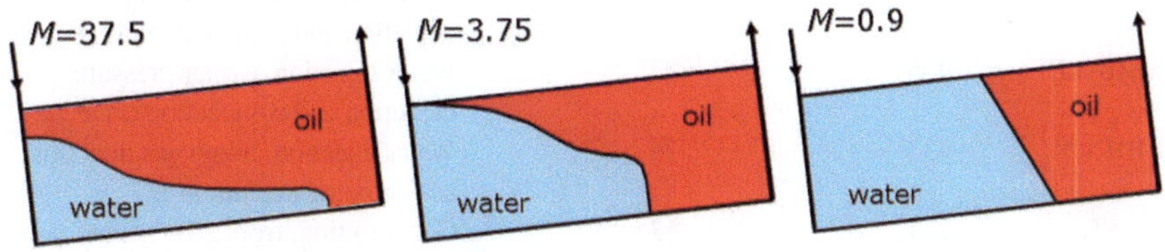

Fig. 2.22 Watersweep with three different mobility ratios

E_A is the *areal sweep efficiency* which reflects how well the reservoir is swept by the waterflood. It depends on the well pattern, the properties of the reservoir fluids (viscosity), the characteristics of the reservoir rocks (permeability), and the rate at which the displacement takes place. E_V is the vertical displacement efficiency which depends on the same properties, but in the vertical flow direction. The *microscopic displacement efficiency* E_D is the recovery rate at *pore level*. It can never be 100 % due to capillary forces which results in a residual oil saturation S_{or}. Oil saturation S_o is defined as the percentage of the reservoir pore volume which is filled by oil.

The volumetric sweep efficiency $E_A \times E_V$ is the fraction of the total reservoir volume contacted by the injected water during the recovery. If the oil is very viscous, its mobility may be low compared to the mobility of the water in which case the water may move around the oil. This results in an unfavorable production situation with a poor sweep efficiency and an early water breakthrough in the production wells, leaving much of the oil behind. Figure 2.22 illustrates this for three different mobility ratios M between water and oil in a slightly inclined reservoir. The highest mobility ratio represents the most viscous oil. On the other hand, a piston shaped displacement with a high volumetric sweep efficiency is the result of a mobility ratio close to unity.

Recovery factors after secondary recovery (e.g. waterflood) range from 20 to 70 %, depending on well strategy. It has been considerably improved during the last decades by effective horizontal drilling and improved sweep efficiency. Table 2.6 ranks the 10

Table 2.6 10 largest oil fields on the Norwegian continental shelf by reserves originally in place (NPD 2013)

Field	STOOIP (mill Sm^3)	Recovery (%)	Main drive mechanism
EKOFISK	1,099	49	Water injection, earlier pressure depletion and compaction drive
STATFJORD	860	66	Pressure depletion in the late phase, earlier water, water alternating gas
TROLL	642	39	Pressure depletion with natural water and gas drive
GULLFAKS	642	61	Water injection. Some gas, and water alternating gas injection
OSEBERG	592	64	Gas injection. Some water alternating gas injection
SNORRE	515	47	Water, gas and water alternating gas injection
ELDFISK	463	29	Water injection, earlier pressure depletion and compaction drive
VALHALL	435	33	Water injection, earlier pressure depletion and compaction drive
HEIDRUN	432	39	Water injection. Some gas injection and pressure depletion
GRANE	229	53	Gas injection, from 2011 water injection and gas reinjection

largest oil fields on the Norwegian continental shelf by reserves originally in place (STOOIP).

2.4 Enhanced Oil Recovery

Enhanced oil recovery (EOR), sometimes called improved oil recovery, is a term used for tertiary recovery methods in which energy is supplied by injecting substances which are not normaly present in the reservoir. The purpose is to increase the recovery factor beyond what can be achieved by primary production methods (pressure depletion) or secondary production (water or natural gas injection), which on the average is of the order of 35 %. Most EOR methods are expensive and on the global scale there has been a clear connection between the oil price and the willingness of the oil companies to start up such measures. Internationally the EOR methods have so far almost exclusively been used on onshore fields, and have created significant values for the owners. Offshore application is still in its infancy, but studies predict that a combination of methods technically has the potential for an increase of 15–25 %.

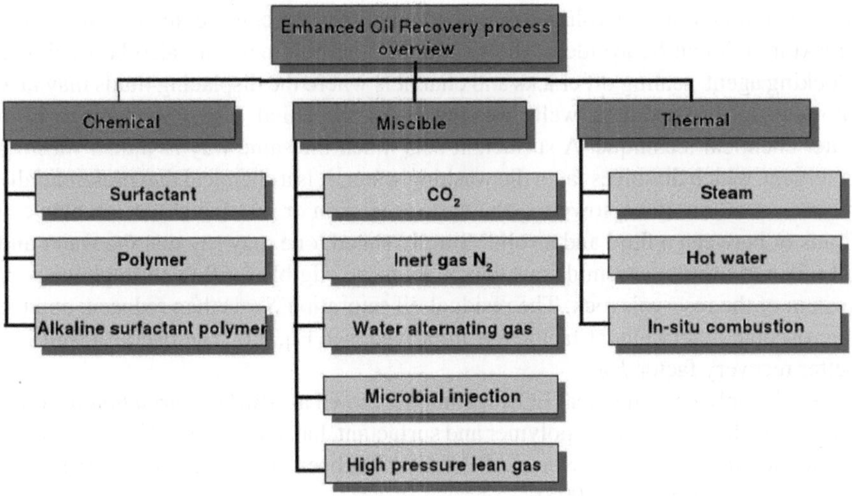

Fig. 2.23 Overview of the most important EOR techniques

Three main categories of EOR exist

- thermal techniques
- chemical techniques
- miscible techniques

The most common EOR techniques are summarized in Fig. 2.23 and described briefly below.

Thermal EOR techniques are used to reduce the viscosity of the oil, thereby improving its mobility, and allowing the oil to be more efficiently displaced to the producing wells. This is the most common EOR technique used on onshore reservoirs with oil gravity less than 20°API. Steam or hot water is injected into the reservoir. This can be done in dedicated injectors (steam or hot water drives) and producing oil from other wells much the same way as with water drives. Alternatively, the steam stimulation and production can be carried out in the same well in a cyclic process called "huff'n puff" in which the steam first soaks the reservoir before it is withdrawn allowing the oil production to take place. A more spectacular method is *in-situ combustion* or fire flooding where heat is generated by igniting a mixture of hydrocarbon gas and oxygen directly in the reservoir. As the fire moves, the burning front pushes ahead a mixture of hot combustion gases, steam and hot water, which in turn reduces oil viscosity and displaces oil toward production wells.

Chemical EOR techniques includes the injection of *polymers* which are long-chained organic molecules with high molecular weight, often one thousand or more. Both biopolymers (e.g. xanthan) or synthetic polymers are used. When injected in water with special chemicals added, the polymer form a *gel* which resembles a thick soup. This increases the viscosity of the displacing fluid, thereby reducing the mobility ratio M between water and oil to a more favorable value, see Fig. 2.22.

Hence, a more efficient volumetric sweep efficiency is achieved and an early water breakthrough can be avoided. The gel formed by polymers can also be used as a blocking agent, sealing off cracks and channels where the displacing fluids may take shortcuts to the producing wells, leaving much oil behind. Use of *surfactants* is an other chemical technique. A surfactant acts much the same way as a dish washing detergent which dissolves fat in the washing water. It is a chemical that preferentially adsorbs at an interface, lowering the surface tension or interfacial tension between fluids or between a fluid and a solid. Simply stated, one may say that the water and oil mix and form microemulsions which are more eligible for flow through the pore system of the reservoir rock. The residual oil saturation S_{or} is then reduced, and the microscopic displacement factor E_D improved, see Eq. 2.6. The overall result is a better recovery factor E_R.

A relatively new chemical EOR technique known as *alkaline surfactant polymer* (ASP) which is a mixture of polymer and surfactant, has successfully been conducted worldwide in recent years. Some ASP floods has been achieving 20 % incremental oil recovery (Zerpa et al. 2005).

Miscible EOR techniques implies the injection of miscible gases into the reservoir. A miscible displacement process maintains reservoir pressure and improves oil displacement because the interfacial tension between oil and water is reduced. Miscible displacement is a major branch of enhanced oil recovery processes. Injected gases include liquefied petroleum gas (LPG), such as propane, methane under high pressure, methane enriched with light hydrocarbons, nitrogen under high pressure, and carbon dioxide (CO_2) under suitable reservoir conditions of temperature and pressure. The fluid most commonly used for miscible displacement is carbon dioxide because it reduces the oil viscosity and is less expensive than liquefied petroleum gas. A special and effective miscible method is called Water-Alternating-Gas (WAG) which has also been applied to offshore fields in the North Sea (NPD 2009). This is a process used mostly in CO_2 floods, whereby water injection and gas injection are carried out alternately for periods of time to provide better sweep efficiency and reduce gas channeling from injector to producer.

EOR techniques require use of large amounts of chemicals and therefore represent a substantial *environmental hazard*. Offshore application creates extraordinary challenges related to the handling of back produced EOR chemicals and discharge to sea. These risks need to be assessed, and handling requirements before discharge must be established. It is necessary not only to use those chemicals which are technically most effective, but at the same time select those which have the lowest risk of environmental damage. The economic impact of implementation of EOR techniques can therefore be tremendous both for the operator and the society.

2.4.1 Oil: Present Use, Resource Considerations and Forecast

Oil is a remarkably valuable resource. Although most of the oil is consumed as a source of energy, refined oil products are key components of many finite products, including plastic, some types of make-up, candles and synthethic fibers used in clothing.

In this book, we focus on oil as a source of energy. Keeping this scope in mind, oil has particular properties that makes it suitable to most applications requiring energy. As of today, oil still contributes to a third of the world primary energy (IEA 2011c). The word "still" in the preceding sentence matters, since most countries have tried to diversify their energy mix in order to smoothen the impact of oil shocks on their economy. As a consequence of this political will and to the emergence of cheaper substitutes for some applications, the share of oil has decreased in the past three decades. The net consumption of oil is however, increasing.

Once the crude oil has been produced, several steps are required before the crude oil can be delivered to the customer in the desired form. Crude oil will first have to be transported from the production site to the refinery. If the production site is located offshore, crude oil is generally transported in tankers or barges. If the production site is located onshore, trucks, trains and pipelines will be used, the choice of which depends on the distance and infrastructure available. Pipelines are economically prefered over short distances. In addition, pipelines will further be used to transport the various oil products from the refinery to the local distribution facilities. Finally, the various oil products, including gasoline, diesel and kerosene are distributed to the customers.

Oil can easily be brought to the point of consumption, stored until use and the energy embedded in oil can be produced at will. These qualities makes oil a fuel of choice in the transportation sector. Consequently, the big chunk of crude oil, or more precisely almost all of the gasoline and diesel obtained from crude oil, is used in the transportation sector.

Oil-derived products are also used to provide heat and some countries (albeit fewer over time because of the emergence of cheap substitutes) burn crude oil directly in order to generate electricity (e.g.: Japan, Moldova, Russia).

The oil consumed globally has increased over time, see Fig. 2.24. In the OECD countries, the demand for oil increased by 10 % between 1980 and 2012 to reach 46.2 million barrel/day[8] in 2012, whereas the demand for oil in non-OECD countries has more than doubled over the same time period (+110 % reaching 44.7 mb/day). Most of the growth in oil consumption over the past three decades is consequently mainly the result of the increasing oil demand in non-OECD countries.

Figure 2.25 shows how the proven reserves of crude oil is distributed over the continents. The total reserves at the end of 2012 were estimated to 1,669 billion barrels (BP 2013), which would suffice to cover our needs for another 50 years at the current rate of extraction. The latter number is however highly uncertain. First, the estimates of crude oil reserves are largely unprecise as they rely on somewhat subjective judgements on the estimated size of the total resource in the ground and on to what fraction of it is economically recoverable (Cassedy and Grossman 1998). Second, proven recoverable reserves have **increased** over the past decades due to the finding of new reserves (mostly in Southern and Central America) and to increasing crude oil prices (recall that proven reserves depend on economical factors too). In

[8] 1 barrel = $0.159\ m^3$

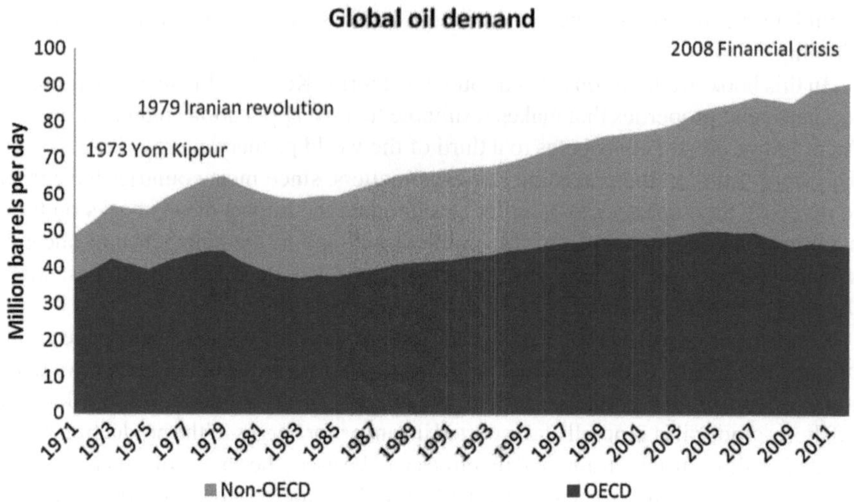

Fig. 2.24 Global oil demand (million barrels per day) (IEA 2013)

facts, global crude oil proven reserves have increased by almost 60 % since 1992 (BP 2013), see Fig. 2.26.

Compared to coal, oil reserves are very unequally spread around the world, with three quarters of the proven reserves located in the OPEC countries.[9] The following four countries: Venezuela (18 %), Saudi Arabia (16 %), Iran (9 %) and Iraq (9 %) together own over half of the world's proven reserves. Other parts of the world are much poorer in terms of how much reserves they own. Such concentrated reserves is considered as a threat by several countries as, in theory at least, some countries have the potential to influence the global oil market.

Countries extracting coal are generally those consuming most of it. This relationship is clearly not true for oil. In fact, there is a clear mismatch between production and consumption of oil. For example, the US consumed twice what they produced in 2012 and China consumed 2.5 times what it produced.

In order to balance supply and demand, some countries must logically produce more than they consume, which is the case of Saudi Arabia, Russia, Canada, Venezuela, Norway and a few others. Such as to balance the system, vast quantities of oil are being traded globally. The main trades are illustrated in Fig. 2.27. It is clear from the that figure that large quantities of oil are exported to the United States, Europe, Japan, China and India.

The future regional demands for oil are uncertain, although the global demand for oil is more than likely to increase. In parallel to the example of Germany, demand for oil in OECD countries is expected to slowly decrease overtime (IEA 2011c) due to increased efficiency, government policies implemented to reduce oil consumption (e.g.: fuel efficiency standards) and to the slow emergence of substitutes to oil (natural

[9] OPEC stands for Organization of the Petroleum Exporting Countries, organization aiming at coordinating the petroleum policies of its Member Countries.

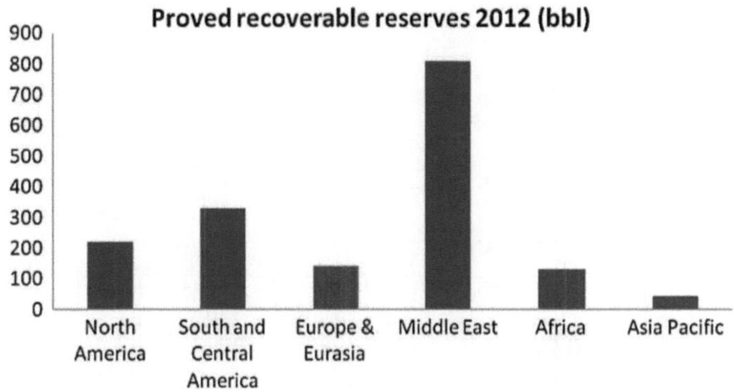

Fig. 2.25 Crude oil proven reserves at the end of 2012 (billion barrels) (BP 2013)

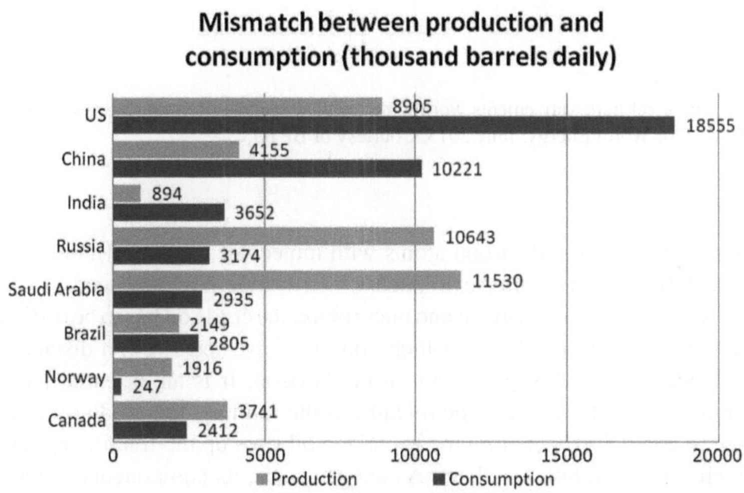

Fig. 2.26 Production versus consumption in 2012 (thousands barrels daily) (BP 2013)

gas, biofuels, hydrogen and electricity). The demand in non-OECD countries is however expected to rise significantly due to an increase in population and to rising standards of living.

Oil Refinery

The crude oil which is recovered from an oil reservoir has a very complicated molecular structure, see Sect. 2.3.1. It ranges from the simplest hydrocarbon molecule which is methane (CH_4) with a molecular weight of 16 to complicated

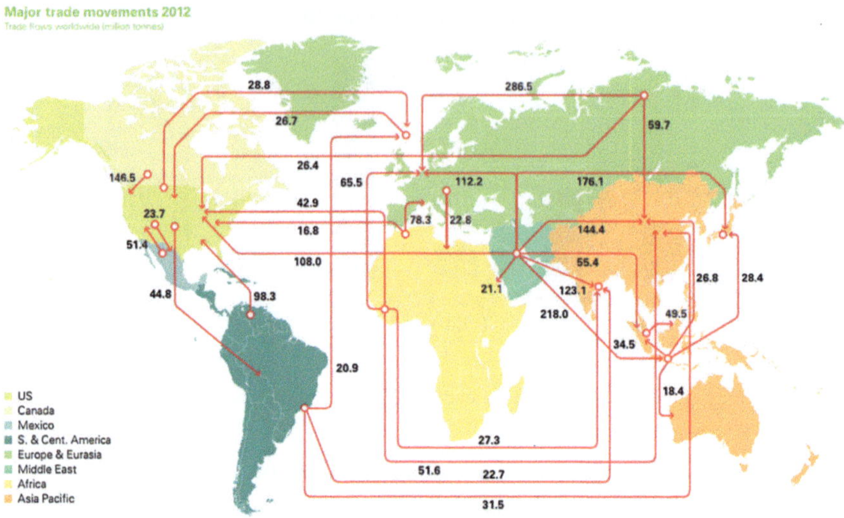

Fig. 2.27 Major oil trade movements worldwide (million tonnes). Reproduced from the BP Statistical Review of World Energy, June 2013. Courtesy of BP p.l.c.

chains of hydrogen and carbon atoms with molecular weight of more than a thousand. In order to obtain the products which a modern society needs, such as gasoline, diesel, fuel oil, asphalt and many more, the crude oil has to be distilled. This is done in an oil refinery, which you may have seen from a distance as tall, slender towers that jut up above the horizon. It is an essential part of the *down stream* handling of petroleum. In the refinery, the crude oil is first heated to about 400 °C before the vaporized oil rises up the fractioning tower through trays with holes in them. As the gas cools, its components condense back into several distinct liquids. Lighter liquids like kerosene and naptha, a product used in chemicals processing, collect near the top of the tower, while heavier ones like lubricants and waxes fall through weirs to trays at the bottom. This is illustrated by the distillation column in Fig. 2.28. The oil refinery is a complex chemical plant and the products from the primary distillation are processed further in other refinery processing units. Demand for gasoline is high, therefore some of the heavier components from the fractioning tower is turned into gasoline by processes called reforming, alkylation and cracking which breaks large hydrocarbon molecules down into smaller ones, making the end product more volatile. Oil refineries are large scale plants, processing about a hundred thousand to several hundred thousand barrels of crude oil a day.

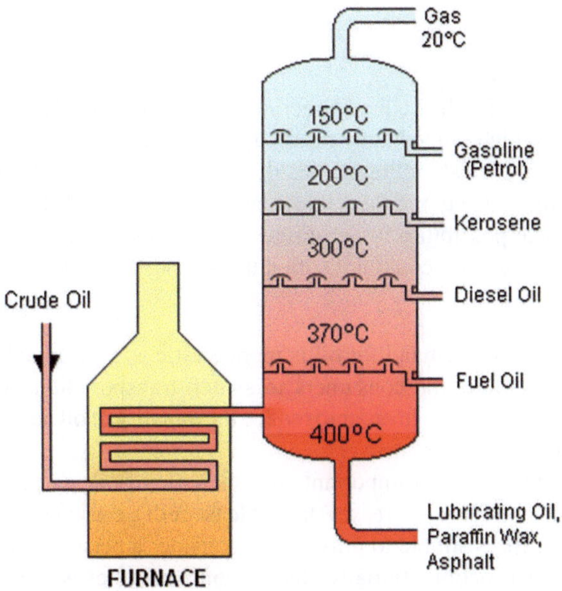

Fig. 2.28 Crude oil is separated into fractions by fractional distillation. The fractions at the *top* of the fractionating column have lower boiling points than the fractions at the *bottom*

2.4.2 Some Remarks on the Economic Characteristics of Oil

Oil is such a key element in today's economy that it is worth spending a few paragraphs emphasizing the particular economic aspects of oil.

Starting with the supply side of oil, it is characterized by a high capital intensity and a high risk (Clo 2000). Once a resource is found and the decision to extract it has been taken, a heavy extraction infrastructure needs to be put in place. Therefore, a large amount of capital will be needed before any oil can be sold. In that sense, extracting oil is a capital intensive entreprise. Even though the technology used to find oil resources has improved significantly in the past, the presence of oil and insights about its economic recoverability will only be known via the act of boring a well. This uncertainty is what makes this industry risky. In addition, high economies of scale and low short term price elasticity are also characteristics of the supply side. The low short term elasticity is due to the rigidity in the petroleum production sector.

On the demand side, economic conditions include very low short term price elasticity, a high income elasticity and cross elasticity. The low short term price elasticity is due to the fact that consumers will often not be able to opt for substitutes if oil prices suddenly increase and the consumers will therefore often be obliged to roughly buy the same quantities of oil even though prices have changed.

The Nigerian example
On January 1st, 2012, the Federal Nigerian Government decided to remove the fuel subsidies which immediately resulted in the doubling of the fuel prices at the pump. Mass protests followed this decision, which created social unrest. Part of the issue is due to the short term elasticities described above. Nigerians did not have the possibility to substitute oil by other sources of energy and therefore the population could not soften the shock.

In the long-term, the demand for oil is more elastic as a sustained price increase might lead, for example, the consumers to switch transportation mode, opt for a different fuel or even relocate in order to reduce the quantity of oil needed to commute between work and home.

Income elasticity is a very important characteristic of oil. As purchasing power increases, the demand for oil increases (true at least until a certain level of purchasing power). This concept is simple to grasp if we consider a person which experiences a big increase in her income. Initially, this person commutes with her bike. As her income increases, she is likely to purchase a car to benefit from better confort (she is now protected from the cold in winter) and save time. This effect does not only happen at the scale of the individual but also at a country scale. As developing countries get richer, their oil consumption increases which results in an increase in the global demand for oil products. Understanding these basic economic concepts is fundamental in order to understand the evolution of the fossil fuel prices, including oil prices.

2.4.3 Natural Gas: Present Use, Resource Considerations and Forecast

In the case of natural gas, almost all of what is currently extracted is eventually transformed into electricity by power plants, utilized by buildings for heating and cooking purposes or used by industries. Natural gas is also increasingly being used as a fuel by combined heat and power plants (CHP) and a very small share (0.15 % in 2007) of the gas extracted is used for transportation. The capital costs necessary to build a gas-fired power plant are low and the construction times short compared to other fossil fuel-based technologies. In addition, natural gas power plants combine a very high operational flexibility, high efficiency levels and a comparatively low carbon emission compared to other fossil fuel energies such as coal. These characteristics naturally makes natural gas attractive to the eyes of many plant operators, investors and governments. Based on these qualities, most of the incremental electricity generating capacity in IEA countries[10] is due to the installation of gas-fired power plants

[10] OECD countries minus Chile, Estonia, Iceland, Israel, Mexico and Slovenia.

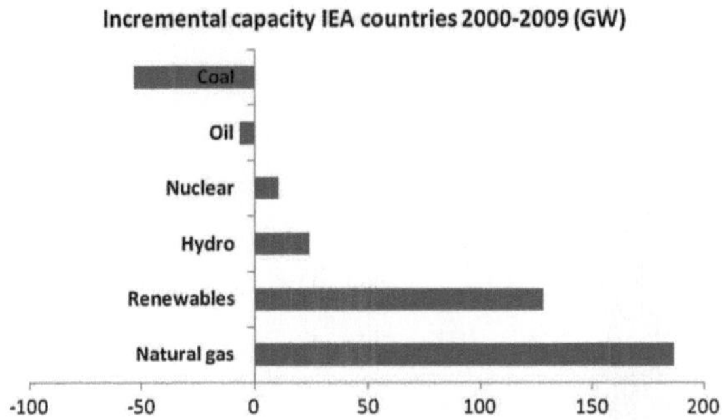

Fig. 2.29 IEA incremental electricity generating capacity in GW, 2000–2009 (IEA 2011b)

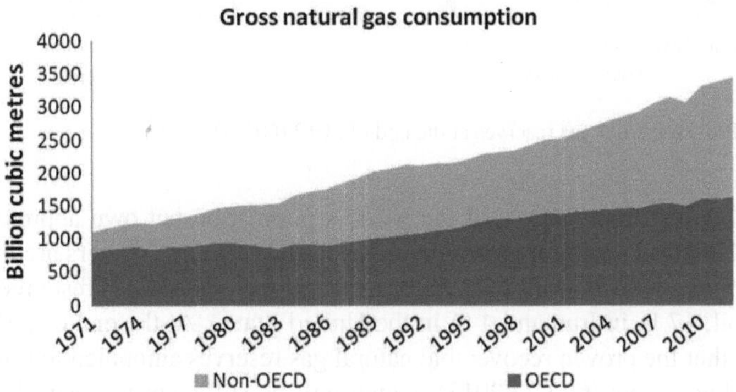

Fig. 2.30 Gross natural gas consumption 1971–2012 (IEA)

as indicated in Fig. 2.29. In the rest of the world, however, coal-fired power plants prevail in accounting for the increase in incremental electricity generating capacity.

Historically, the transformation of natural gas into electricity has been gaining in importance and the natural gas consumption has risen steadily over the past decades. Yet, non-negligible quantities of that precious resource are being flared (burned in the open air) by oil extracting companies focusing solely on crude oil due to cost reasons.

Similar to the case of crude oil, consumers of natural gas are not necessarily the countries extracting it. In 2012, the main natural gas producers were the United States (19.8 % of the world production), the Russian Federation (19.1 %), Qatar, Iran and Canada (each with 4.6 %). That year, the main consumers were the US (20.9 %), the Russian Federation (13.7 %), Iran (4.5 %), Japan (3.8 %) and Canada (2.9 %) (OECD 2011). Figure 2.30 shows the evolution of the natural gas consumption and production for OECD and non-OECD countries since 1971. The 2008 financial crisis is clearly visible as it negatively affected the global natural gas consumption, although the effect of this crisis were short-lived.

Natural gas reserves are not evenly spread around the planet and some countries control a big share of these reserves, see Fig. 2.31. Such an example is Qatar, which

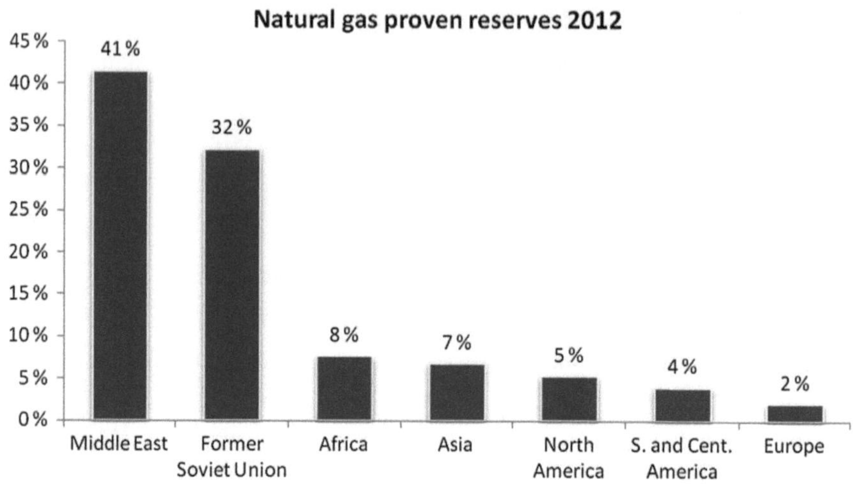

Fig. 2.31 Proven natural gas reserves at the end of 2012 (OECD 2013)

accounts for less than 0.02 % of the world's population but own approximatively 13 % of the world's natural gas reserves. Other big resource owners are the former Soviet Union countries with over 30 % of the global natural gas resources located in their soil, 17 % in Iran and 4 % in the United States. At the end of 2012, it was estimated that the proven recoverable natural gas reserves amounted to roughly 192 trillion cubic meters (OECD 2013), which will be sufficient to cover the demand for the next 60 years at the 2012 rate of extraction. However, due to its flexibility, cost and environmental benefit, it is expected that the demand for natural gas will increase significantly in the near future.

Natural gas is not available everywhere and this commodity thus needs to be transported to where it is needed, the cost of which varies with respect to the distance, the scale and the transportation mode chosen. Liquefied natural gas (LNG), offshore and onshore pipeline are the existing technologies used to transport natural gas. Figure 2.32 illustrates the economics of these technologies.

LNG requires that the natural gas is first liquefied before transportation and regasified before use, which comes at a cost. Yet, this incremental cost rises slower than the incremental cost of building longer onshore and especially offshore pipelines (Jensen 2007), therefore LNG transportation will be priviledged over long distances.

Europe gets its natural gas mainly via the use of pipelines from the North Sea, Algeria and from Russia. Supply and demand being inflexible in the short term (recall the concepts of elasticities discussed previously), disruption of natural gas from any of these places can quickly threathen the continuous natural gas supply to central European countries. For example, the January 2009 natural gas supply disruption of the gas transiting through Ukraine for political reasons affected 18 European countries.

The IEA expects the demand for natural gas to increase by 30–65 % by 2035 (IEA 2011c), mostly because of increasing natural gas demand from non-OECD countries and to the substitution of fossil-fuels by natural gas in the OECD countries. An

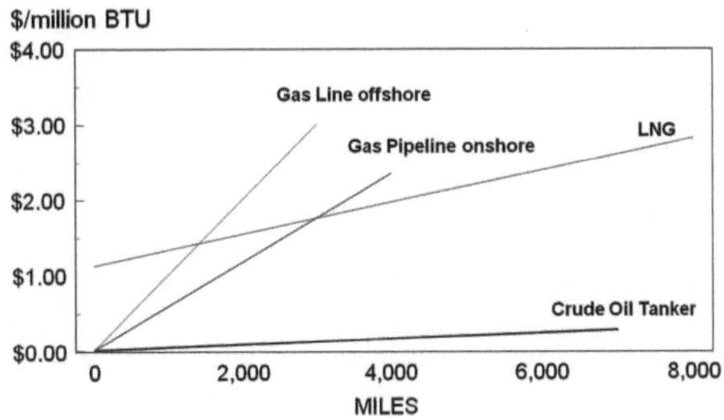

Fig. 2.32 Gas transportation costs in 2002 in USD/MMBTU (Jensen 2007)

increase of that magnitude means that in the absence of new natural gas field discoveries, the proven reserves will not last until the end of the century. However, the development of unconventional gas resources and shale gas principally, can mean that natural gas reserves will be sufficient for a much longer period (more about this in Sect. 2.7).

2.4.4 Cost

2.4.4.1 Resource Cost of Oil and Natural Gas

The oil markets are integrated to such a point that major events impacting oil producing countries affect the crude oil prices worldwide. As Fig. 2.33 shows, oil prices have been relatively stable in real terms between 1986 and 2003 as growth in consumption was matched by growth in production and technological improvement. Shocks on prices disappeared with either the cause of the shock or because other oil producers managed to increase their production. Between 2003 and 2013, oil prices have fluctuated more violently. First, the addition of the Iraq war with growth in crude oil demand from rapidly developing countries and the failure to increase production at the same rate as the demand, generated an energy crisis where the theme of peak oil became popular. The 2008 financial crisis resulted in a violent market correction which was short lived as global economy started its recovery. The Arab Spring pressured oil prices upward since crude oil production was largely disrupted in that country and because of the importance of Libya in the oil production market. The strength of the shock is, once again, due mostly to the very low price elasticity of the oil commodity and to the quasi-absence of substitutes.

Figure 2.34 illustrates the monthly natural gas prices in real terms for the United States and Europe between 1970 and 2013. Comparing Figs. 2.33 and 2.34 shows that both resources have prices that tend to be highly correlated: as the oil price goes up or down, the price of natural gas follows the same trend.

Formally, natural gas prices are subjects to four main characteristics. First, supply and demand levels naturally play an important role in defining natural gas prices. The

Fig. 2.33 Crude oil prices in 2008 Euro (World Bank 2013)

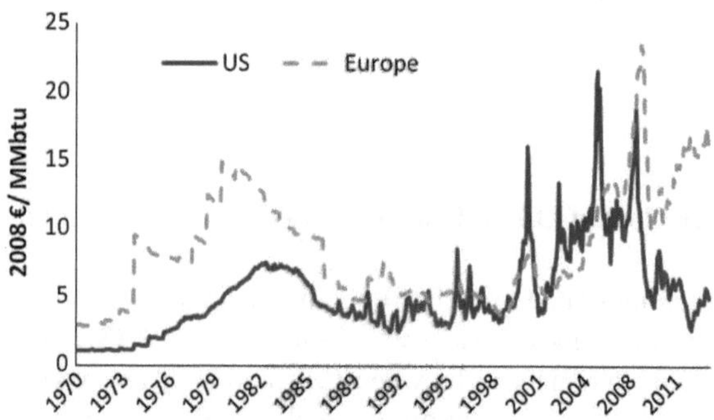

Fig. 2.34 Natural gas prices in Euro^{2008}/MMBtu (World Bank 2013)

role played by major events on demand and supply should be clear after what has been written previously on coal and oil. For example, at least part of the drop in prices in the second part of 2008 can be attributed to decreasing industrial pressures on demand from the industry due to the financial crisis. Second, regional natural gas markets are increasingly interlinked. The various regional markets used to be more independent from each other, however, with the creation of numerous new LNG routes due to the emergence of that technology, regional markets are better interlinked (IEA 2009). Note however the divergence in prices between US and European gas prices since 2009. The diverging trends between the US and European natural gas prices is mainly due to the production of unconventional gas resources in the US, e.g.: shale gas, while this exploitation is often limited in Europe because of environmental concerns (e.g. France). The diverging trend is an indicator that the export capacity is insufficient (for now) in the US for the market prices to converge totally. In fact, before the emergence of the shale gas industry in the US, many LNG import stations have been built in prediction of a large shortfall in the natural gas production in the US. These stations cannot be used for export unless extensive transformation is taking place, which means that divergence in natural gas prices between the US and Europe will be possible until appropriate trade facilities are put in place. Nonetheless, regional natural gas price patterns are likely to converge even more in the future and that regional shocks will be short-lived as natural gas can be supplied from or to another region.

Thirdly, long-term natural gas delivery contracts are often linked to the evolution of oil prices (IEA 2009) explaining the close link between the prices of the two resources. Lastly, the availability of substitutes to natural gas in the end-market impacts the natural gas price evolution, limiting the amplitude of shocks on natural gas prices.

More than oil, natural gas is used to generate electricity and the cost of fuel accounts for the big part of its levelized cost of electricity (over 60 %). Due to this high share, gas-fired power plants are the most sensitive to fuel price variation, even more so than coal. Therefore, it is especially important for the investor to properly estimate the cost of natural gas over the lifetime of the power plant in order to get proper estimates for the cost per kWh. The variations in natural gas prices make such estimates impossible. Instead, investors use a steady fuel cost increasing rate based on what the theory predicts.

2.4.4.2 Basic Cost of Natural Gas Power Plants

The numbers provided in this section are based on gas-fired combined cycle turbine (CCGT) power plants as this type of plant is particularly popular today.

The capital costs of a CCGT power plant cover basic elements such as those related to the turbine, the cooling systems for the steam condenser, water treatment and maintenance facilities (Council 2002) and the construction costs including materials, engineering and construction management. The capital cost of a CCGT power plant will depend on the location with regard to the electrical grid and the natural gas network. In addition, the location will influence the technology chosen as the absence of a large and stable source of water to cool the steam will force the developer to use a closed-loop system, which results in lower efficiency levels and also in higher capital costs. Regulations can impact the capital costs as more stringent pollution control requirements will eventually result in higher capital costs. Two years are usually sufficient to commission a CCGT power plant.

The capital costs of CCGT power plants reach Euro 350–1000 Euro/kW of installed capacity. Assuming a plant economic life of 30 years, these capital costs account for a small share (15–30 %) of the levelized costs of electricity compared to other technologies.

2.4.4.3 Electricity Generation Cost of Natural Gas Power

Natural gas is ‘cheap’ compared to other energy technologies. Capital costs are low, construction times short and efficiency levels high. Operation and maintenance work

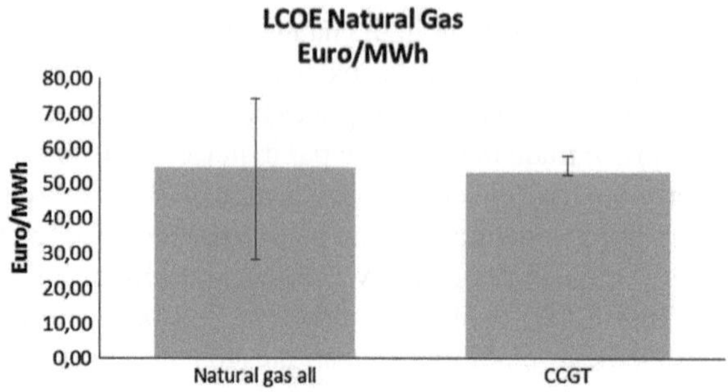

Fig. 2.35 Levelized cost of electricity: natural gas

Table 2.7 Levelized cost of electricity for natural-gas-fueled power plants

	All natural gas	CCGT
LCOE (Euro/MWh)	28–74	52–58
Overnight cost (%)	7–31	14–31
O&M costs (%)	2–11	2–11
Fuel costs (%)	61–87	61–80

is as necessary for natural gas power plants as for any other technology. However, these costs are particularly low for natural gas (Eurocents 0.1–0.5 Eurocents/kWh).

Summing up the levelized capital costs, O&M costs and fuel costs result in a LCOE of Eurocents 2.8–7.4 Eurocents/kW of electricity generated from natural gas for the minimum and maximum cases respectively. The maximum case is based on a plant in Japan, where the cost of fuel (LNG) is particularly high since it has to be imported. Because fuel costs account for the big share of the LCOE, the smallest price variation will be felt by the plant owner. Therefore, the exact LCOE of natural gas cannot be precisely known in advance as it relies heavily on hypotheses on future natural gas prices.

2.5 Carbon Capture and Storage

Three different classes of proposals to mitigate global warming resulting from increased concentration of atmospheric CO_2 due to fossil fuel have been raised:

1. Replace fossil fuels by energy sources which do not release any significant amount of CO_2 (e.g.: nuclear or renewables).
2. Continue using fossil fuels and at the same time, capture and store significant amounts of CO_2 and consequently reduce the total emissions into the atmosphere.
3. Apply various types of *geoengineering*, i.e. to introduce enterprises on ground, in the atmosphere or in space with the aim of counteracting the effect of increased CO_2 levels.

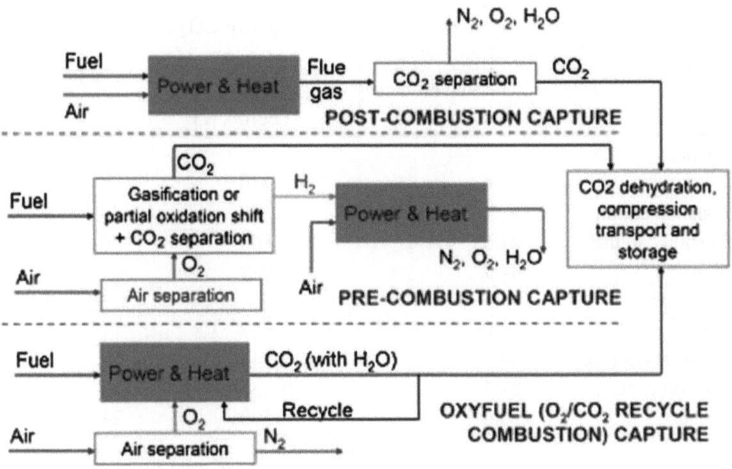

Fig. 2.36 Overview of the main CO_2 capture technologies (Gibbins and Chalmers 2008)

A range of suggestions exist under item 3, from placing mirrors outside the atmosphere to reflect sunlight to painting buildings and large areas on the ground in white such as to increase the albedo. We will not describe these issues any further, as they are considered outside the scope of this book. The alternative energy sources to fossil fuels will be described in the following chapters. In this section, we describe the carbon capture and storage (CCS) option.

The vision behind CCS is to allow for continued use of fossil fuels without contributing to global warming. One way of achieving this vision would be to collect the CO_2 and pump it back into the oil and gas reservoirs. This process has been carried out at a small scale with an industrial motivation for many years. Already in the 1970', the injection of CO_2 into oil fields in order to increase oil recovery was initiated and the idea of using depleted oil reservoirs as a store for human CO_2 production possibly has its origin in these projects.

It is important to realise that we have no realistic solutions for CCS in the transportation sector. CCS is therefore only a valid solution to CO_2 emission from power plants, in practice from coal-fueled and natural gas-fired power plants.

A range of technologies can capture the CO_2 before it is released. These are summarised in Fig. 2.36.

The figure shows the different technologies possible for CO_2 capture. Very briefly, there are two alternatives: Either the fuel is transformed in such a way that CO_2 can be extracted before (pre-combustion) or after (post-combustion) the combustion process. Pre-combustion implies that coal is gasified and turned into several new components such as CO_2, CO and H_2. The H_2 gas can be separated and used for energy production while the CO_2 is captured. In post-combustion processes the coal or hydrocarbons are first burned resulting in CO_2 as end products together with

Table 2.8 Thermal efficiency, capital costs and costs to remove CO_2 for various coal and gas at various technologies (Gibbins and Chalmers 2008)

Technology	Thermal efficiency (% LHV)	Capital cost (Euro/kW)	Cost of CO_2 avoided (Euro/tCO_2)
Gas-fired plants			
No capture	55.6	340	
Post-combustion capture	47.4	590	40
Pre-combustion capture	41.5	800	76
Oxy-combustion	44.7	1,038	70
Coal-fired plants			
No capture	44.0	957	
Post-combustion capture	34.8	1,344	23
Pre-combustion capture	31.5	1,235	16
Oxy-combustion	35.4	1,500	24

water H_2O and other gases. CO_2 may then be extracted by sending the end products through a fluid which binds CO_2. Figure 2.36 also shows a combined version of these two techniques, the so-called oxyfuel process: Here natural air is replaced with pure oxygen in the combustion process which results in a much cleaner exhaust containing only CO_2 and H_2O. The water may now easily be condensed allowing for a cheaper collection of CO_2. However, the drawback with this method is the extra expense to replace air with clean O_2 before combustion.

It is clear that the process of removing CO_2 requires energy. It is easy to realise that carbon capture reduces the efficiency of the power plant, since this energy is naturally provided by the plant itself. We see in Table 2.8 that post-combustion is the best option in this case, and imply a reduction factor of about 20 %. In addition we see that capital costs also increase since the capture technology has to be installed as well. The LCOE in total can reasonably be expected to increase with about 20 % for the best capture technology. This adds up to a total cost per captured tonne CO_2 in the range from Euro 20 (coal) to 40–70 (gas). A massive program to install such technologies will likely slightly reduce costs on the capital side. Energy penalties and reduced efficiency are determined by laws of nature and cannot significantly reduced.

The storage problem is in practice more challenging than capture, and will also require a considerable amount of energy. In order to contribute significantly to climate change mitigation, the amount of CO_2 to be stored is of the order a few to ten Gt (Gigatonnes) per year since the present annual emissions are about 30 Gt per year. We have no experience in building such programs and, for comparison, the total amount of CO_2 captured so far during the last 30 years since the idea was conceived amounts to some 10 Mt, i.e. one thousandth of what we need to capture anually! When 10 Gt atmospheric CO_2 is compressed and cooled into a liquid, it would be take as much space as pilling up 4,815 Pentagon, the famous headquarters of the US Department of Defense:

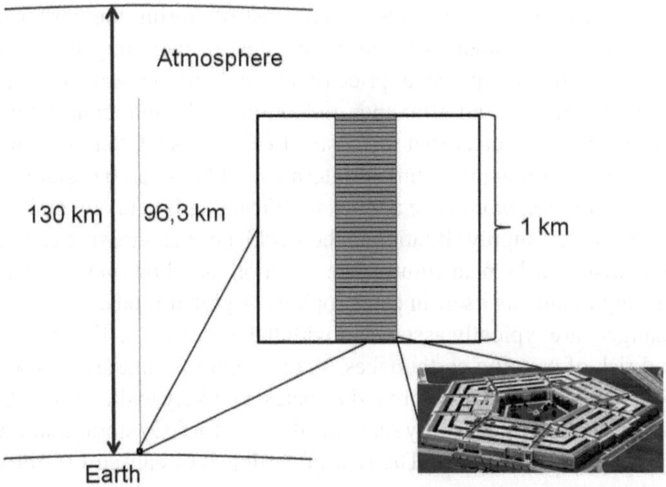

Fig. 2.37 10 Gt of liquid CO_2 is in order of magnitude similar to pilling up nearly 5,000 Pentagon (each side of the Pentagon is approximatively 280 m, and the height of the building is roughly 20 m)

The following storage possibilities are known:

- Ocean storage. The ocean is a gigantic CO_2 storage and contains, as seen from Chap. 1, more than 1,000 times the amount of annual anthropogenic CO_2 emission. By releasing CO_2 at about 1 km depth, it is therefore in principle possible not to change the CO_2 significantly on a grand scale. The storage capacity of the oceans are then in principle sufficient, but the price factor of pumping CO_2 down to 1 km which require a compression of liquid CO_2 about a 100 times represents the main barrier. Releasing CO_2 at shallow water would on the other hand be cheap, but in that case it would mix with the oceans biosphere and potentially damage it.
- Geological storage. Depleted oil and gas reservoirs are ideal for "returning" CO_2. Extraction wells can be changed into injection wells. That world storage capacity is estimated to be in the order of 100–1,000 GtC. This amount is increased at least by a factor of two by adding deep saline aquifers and coal seams. Realistically this option then allow for a annual uptake of 5 GtC for more than 50 years which is considered a sufficient time window for continued use of fossil fuels without global heating allowing for new energy sources to be implemented. However, storing liquid CO_2 in depleted reservoirs or other underground voids also requires energy because they are usually filled with water which has to be displaced.
- Biological storage such as reforestation would also contribute to an uptake of CO_2. Large areas of countrysize dimension needs to be planted on land. Other techniques based on plancton, algae etc in the sea have also been suggested. The total capacity is estimated to be an order of magnitude smaller than for geological storage sites.

As a summary we see that sufficient storage capacity exists in principle. The cost of any of these storage technologies is uncertain, and varies from as little as from

1 Euro per tonne to 70 Euro per tonne. The cost of storing the carbon dioxide is simply anything from a small fraction off the cost of capturing it to twice nearly twice that cost. Setting the physical price of storage[11] to the same level as the price of capturing CO_2 implies an efficiency reduction of thermal coal plants of about 40 %. Thus, it should be clear that the costs of CCS are such that it is unrealistic to foresee a global implementation unless international laws can be established which prohibits CO_2 emission or enforce a more significant carbon tax on emissions. With an unbalanced energy supply situation in the world, no such measure can be foreseen to become realistic in the near future. The situation may however rapidly change if global warming manifests itself in catastrophies or global food crises.

Some dangers are typically associated with the use of CCS. These are the risk of leakage and risk of creating earthquakes. In the event of a massive carbon leakage, people in contact with the gas before it dissipates are likely to die. The risk of deadly leakage can be lowered by carefully selecting the sites for CO_2 storage and suppressed by storing the gas off the coast. The risk of earthquakes caused by storing carbon dioxide underground results from a change in pressures in the ground, in the same way that extracting oil and gas can create an earthquake. The magnitude of potential earthquakes is expected to be tiny. However, it is generally not a good idea to cause them at the first place because it can leads to fractures in the sealing layer that prevents the carbon dioxide from escaping. The creation of earthquakes depend on a variety of parameters, with the type of sealing layer and the state of stress of the material being amongst them. Carefully selecting and monitoring the storage sites can lower this risk.

Application: Technology Centres Mongstad, Norway
The Technology Centre Mongstad located in Norway is one of the major research centre for carbon capture and storage. A variety of post combustion CO_2 capture technologies are being tested and improved exploiting the exhaust gases from a natural gas-fired power plant built at this centre.

About 800 million Euros had already been poored into the project by mid-2012 and the plant managed to capture 22,000 tonnes of CO_2 that year. The captured CO_2 was eventually released back into the atmosphere in the absence of a storage facility. Once fully operational, the plant should have been able to capture a Mt CO_2 annually. In the fall of 2013 the new Norwegian goverment decided to severely reduced its support to the centre for economical reasons.

[11] The physcial price is the loss in efficiency here.

Fig. 2.38 Technology Centre Mongstad, Norway (courtesy TCM ©)

2.6 Peak Oil

How long can the oil supply satisfy the oil demand? This question depends on at least three factors. The first factor concerns the total amount of the resource which is found and can be extracted. Apart from the obvious fact that the quantity has to be finite, there is no exact public knowledge of how large our remaining petroleum resources exactly are. In addition, the constant ongoing search for oil results in a almost continuous discovery of new oil fields. The present discovery rate is about 10 Gb (Gigabarrels or 10^9 oil barrels) per year, which is significantly lower than the annual consumption of about 30 Gb.

A common term in the industry is to consider "2P" reserves as the best estimate for the remaining commercial resource base for conventional oil. 2P means the already proven reserves and the unproven reserves which are "likely" to be found with a probability larger than 50 %. The global world estimate of the 2P reserve has been evaluated to be about 1,000 Gb. The uncertainty is about ±300 Gb. Up till now (2013), humans have already used about 1,100 Gb (Aleklett et al. 2012) which shows that as compared to a bottle of Coke, we have now most likely emptied more than half of it. Also, compared to the present annual consumption of oil we see that the resource is emptied in about 30 years from now if the production is kept constant.

In addition to the uncertainty in the resource base there are other factors which affect our estimate. The first is increased demand, through population growth and increased industrialization of large countries like India and China. A second problem is related two what today is known as the Hubbert "Peak Oil Theory" (Hubbert 1956). It says that the production from a finite resource can only increase for a limited time until a maximum production is achieved. This time is called "Peak oil", t_{peak}. After "Peak Oil" the production will decline no matter how much the demand would be—unless revolutionary new production technologies are discovered. The only way to meet an increased demand after peak oil for a given resource is to find additional resources. Regarding conventional oil this is at present more unlikely than likely: Almost half of the oil resources exist in so-called giant oil fields which were discovered more than 50 years ago. Continuous search in increasingly more harsh environments (arctics, deep oceans) have so far not changed the almost constant and slowly decaying curve of new discoveries. In 2011 three new Norwegian oil fields were found in the North Sea and in the Barents Sea which were reported as significant and large. However, the estimated total amount of oil within these fields will only be able to cover the present global oil consumption for about a month!

The Hubbert curve can be derived from relatively simple physical models of production from a reservoir of finite size (Helseth 2012). We will consider the total accumulated produced volume $V(t)$ of recoverable oil from a reservoir. The produced volume per time, dV/dt will be given by the product of the total area of producing units $A(t)$ and the relative volume fraction of remaining producable oil $S(t)$ and the flow velocity, $u(t)$ of the oil through the producing area (a number of production pipelines with total cross sectional area $A(t)$),

$$\frac{dV}{dt} = S(t)A(t)u(t) \tag{2.7}$$

By injecting fluids through some injection wells it is possible to keep the pressure within the reservoir almost constant. Then the velocity, $u(t)$, through the pipelines is almost constant. The flow area is determined by the number of production wells. Such are expensive to drill in general and a first order approximation is that the number of production wells at any time is proportional to the total volume of produced oil, $N(t) = N_{\max} V(t)/V_{\text{tot}}$. Here $N_{\max}$ is the maximum number of wells bored when $V(t) = V_{\text{tot}}$, which is the total amount of oil originally in the reservoir, often called STOOIP (Stock Tank OIL Originally In Place). This implies that $A(t) \propto V(t)$. The relative fraction of remaining oil in the reservoir will also be proportional to $V(t)$ through $S(t) = S_0[1-V(t)/V_{\max}]$, were S_0 is the initial fraction of oil in the reservoir. Putting all this together we obtain the following equation for the production speed,

$$\frac{dV}{dt} = K_p V(t)\left(1 - \frac{V(t)}{V_{tot}}\right) = K_p\left(V(t) - \frac{V^2(t)}{V_{tot}}\right), \tag{2.8}$$

where all the constants have been collected in a factor K_p which has the unit inverse time. The condition for the production peak velocity is $d/dt(dV/dt)_{t_{peak}} = 0$ which gives $V_{peak} = V_{tot}/2$ and therefore

$$dV/dt|_{t_{peak}} = K_p \frac{V_{tot}}{4}, \tag{2.9}$$

This expresses a linear relationship between the total oil produced V_{tot} and the maximum production speed. The solution of the differential Eq. (2.8) is satisfied by the following function $V(t)$ for the accumulated production,

$$V(t) = \frac{V_{tot}}{1 + e^{-K_p(t-t_{peak})}}. \tag{2.10}$$

This relationship can be differentiated to give the instant production speed,

$$\frac{dV}{dt} = \frac{K_p V_{tot} e^{-K_p(t-t_{peak})}}{\left(1 + e^{-K_p(t-t_{peak})}\right)^2} \tag{2.11}$$

This curve has its production maximum at t_{peak} and falls towards zero (no production) when $t \gg t_{peak}$.[12] The curve can be fitted pretty well to the world total oil production (Aleklett et al. 2012) as well as to individual reservoirs. For example, by fitting the parameters V_{tot} and t_{peak} to official figures of Norwegian oil production we can plot the Hubbert curve [using Eq. (2.11)] in the figure below. The data are taken from the resource report of The Norwegian Petroleum Directorate (NPD 2013).

The accumulated oil production by the end of 2012 from existing producing Norwegian fields is about 26 Gb. Figure 2.39 shows the accumulated production (top) and the annual equation using Eq. (2.11) and with using t_{peak} and V_{tot} from the production data (NPD 2013) and using K_p as a fitting parameter. A pretty good agreement is noted, and we see that Norwegian oil production peaked around year 2000 and is now on a rapid decline.

Resources is a collective term for recoverable petroleum volumes (oil, gas and condensates). The resources are classified into the following categories: decided by the licensees or approved by the authorities for development (reserves), volumes dependent on clarification and decisions (contingent resources) and volumes expected to be discovered in the future (undiscovered resources). The main categories are thus reserves, contingent resources and undiscovered resources. The Norwegian petroleum resources are approximately 85 Gb o.e. (oil equivalents). Of this, a total of 37.7 Gb o.e. have been sold and delivered (oil, gas and condensates), which corresponds to 44 % of the total resources. The total remaining recoverable resources amount to 48 Gb o.e. Of this, 31 Gb o.e. have been discovered, while the estimate for undiscovered resources is approximately 17 Gb o.e., which is of course very

[12] There is an unimportant inconsistency at $t = 0$ with (Eq. 2.10) since $V(0) \neq 0$. It can be removed by subtracting the constant $V(0)$ from $V(t)$, i.e.. $V(t) \longrightarrow V(t) - V(0)$.

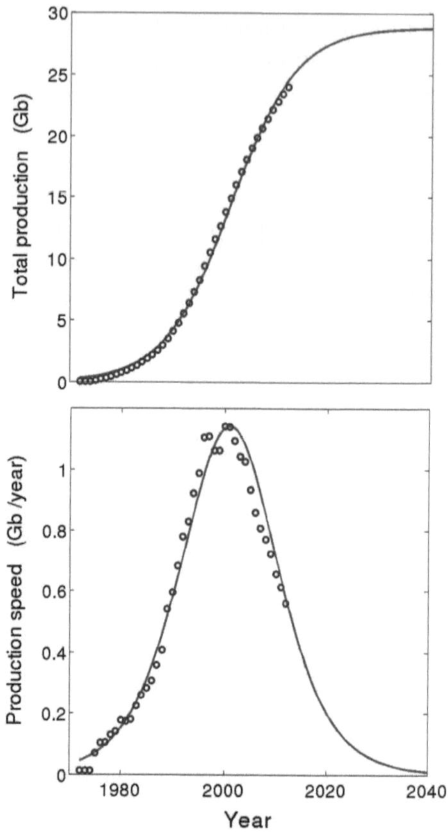

Fig. 2.39 Norwegian oil production (NPD 2013) in giga barrels since 1970 shown as circles compared to the Hubbert curve, Eqs. (2.10) (*top*), (2.11) (*below*)

uncertain. The contingent and undiscovered oil resources are naturally not included in the Hubert analysis presented in Fig. 2.39. When or if they are opened for new production they will have their own production history and Hubbert curve which must be added to the existing one. Only a giant discovery of the magnitude of, or larger than, the initial size of the resources can maintain and increase the production. It is more likely that new discoveries as well as improved recovery technologies only will add up to an extra shoulder of the production history, and expand the Norwegian oil production era for another 10–20 years.

According to the works of Aleklett et al. (2012), the world production as a hole may already have passed peak oil. Others estimate that it may not occur in another 10–20 years which in any case is very soon. The implications will become dramatic: It means that the oil production, very soon cannot meet the increasing demand. This may again lead to extreme oil prices, global political instability, economic depressions and even war. On the other hand it paves the way for commercial utilization of new energy sources, which may be unconventional fossil fuels in combination with increased use of coal and/or new and increased nuclear energy production and/or renewables.

What is meant by unconventional fossil fuel is treated in the following section, while nuclear energy and renewables follows in the forthcoming chapters.

2.7 Non Conventional Fossil Energy Sources

2.7.1 Basic Definition

Conventional oil and gas are produced by drilling into reservoirs, onshore or offshore, where the fluids have accumulated in traps formed by impermeable rock stratas, see Fig. 2.14. This technology has been around and refined over the last 100 years or more. It has been described to some detail in the previous sections of this book. Since the production from these conventional fossil fuel resources inevitably has to decline in the near future, oil industries and governments across the globe are investing unconventional oil and gas resources. The most important of these are

- tar sands or extra heavy oils
- shale gas
- shale oil
- coal conversion
- methane hydrates

They represent enormous energy resources but also involve huge technical and environmental problems.

2.7.2 Tar Sands or Extra Heavy Oils

Heavy oil are highly viscous, “cold syrup” like hydrocarbons which are formed much the same way as the conventional low viscous oil. The oil sands or tar sands are loose sand or partially consolidated sandstone containing naturally occurring mixtures of sand, clay, and water, saturated with a dense and extremely viscous form of petroleum technically referred to as *bitumen.* In the Athabasca sands in Alberta, Canada (see Fig. 2.41) there are very large amounts of bitumen covered by little overburden, making surface mining the most efficient method of extracting it. The overburden consists of water-laden muskeg (peat bog) over top of clay and barren sand. The oil sands themselves are typically 40–60 m deep, sitting on top of flat limestone rock.

Due to the high viscosity, heavy oil is more expensive to extract than conventional oil. The bitumen in tar sands cannot be pumped from the ground in its natural state. Therefore *surface mining*, requiring large areas, has been the preferred extraction technique. After excavation, hot water and caustic soda (NaOH) is added to the sand, and the resulting slurry is piped to the extraction plant where it is agitated and the oil skimmed from the top. Provided that the water chemistry is appropriate to allow bitumen to separate from sand and clay, the combination of hot water and agitation

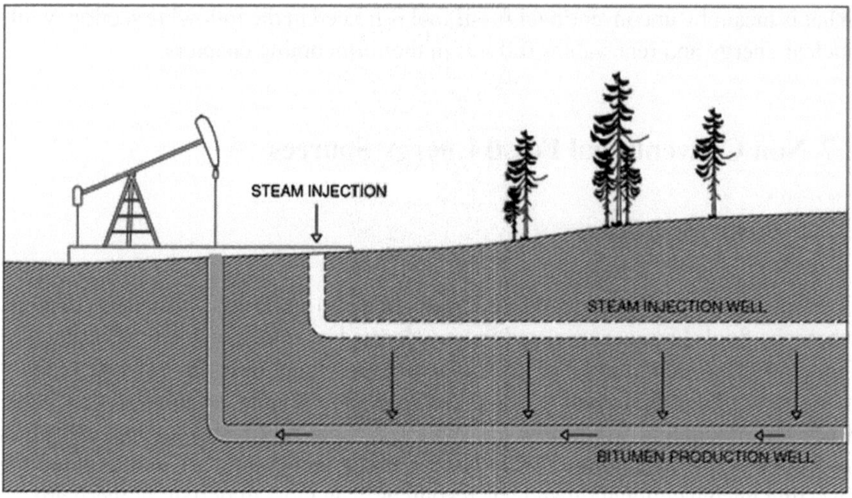

Fig. 2.40 Steam Assisted Gravity Drainage

releases bitumen from the oil sand, and allows small air bubbles to attach to the bitumen droplets. The bitumen froth floats to the top of separation vessels, and is further treated to remove residual water and fine solids. About two tonnes of oil sands are required to produce one barrel (ca. 1/8 of a tonne) of oil. Originally, roughly 75 % of the bitumen was recovered from the sand but today extraction plants recover well over 90 % of the bitumen in the sand.

Several techniques which restrain from occupying large surface land areas have also been employed, e.g. *Cyclic Steam Stimulation* (CSS). In this method, the well is put through cycles of steam injection, soak, and oil production. First, steam is injected into a well at a temperature of 300–340 °C for a period of weeks to months. Then, the well is allowed to sit for days to weeks to allow heat to soak into the formation. Later, the hot oil is pumped out of the well for a period of weeks or months.

In the *Steam assisted gravity drainage* (SAGD) process, two parallel horizontal oil wells are drilled into the formation, one about 4–6 m above the other (Deutsch and McLennan 2005). The upper well injects steam, and the lower one collects the heated crude oil or bitumen that flows out of the formation, along with water from the condensation of injected steam (see Fig. 2.40). The basis of the process are that the injected steam forms a "steam chamber" that grows vertically and horizontally in the formation. The heat from the steam reduces the viscosity of the heavy crude oil or bitumen which allows it to flow down into the lower wellbore. SAGD has proved to be a major breakthrough in production technology since it is cheaper than CSS, allows very high oil production rates, and recovers up to 60 % of the oil in place.

Several more exotic techniques have also been tried at the tar sand fields. For example using solvent instead of steam to separate the bitumen from the sand, or an in-situ combustion process which ignites oil in the reservoir and creates a vertical wall of fire moving from the "toe" of the horizontal well toward the "heel", which

Fig. 2.41 The Athabasca tar sands in Alberta, Canada

burns the heavier oil components and upgrades some of the heavy bitumen into lighter oil right in the formation.

The cost of producing one barrel of tar sands or extra heavy oil amounts to between 30 and 60 $Euro^{2008}$/bl of oil equivalent, to which transport costs must be added. With current oil prices (70 $Euro^{2008}$/bl in September 2013), it makes economic sense to extract this resource. It has been estimated (perhaps conservatively) that, given the existing technology and current oil prices, 1–1.5 trillion barrels of tar sands and extra heavy oil can be recovered economically (IEA 2013b), most of them being located in Canada and Venezuela.

All production methods of oil from tar sand requires large amounts of energy, chemicals and water. It also releases considerable amounts of CO_2 into the atmosphere. An upheaval of large land areas and substantial pollution is often the result. Needless to say there are strong environmental concerns regarding extraction of energy from these resources. However, with increasing oil prices, tar sands and extra heavy oil are considered important from an industrial perspective, especially since the resource base is significant.

Application: Canadian oil sands.
From an economical perspective, the cost of extracting tar sand depends on the quality of the reservoir, the location of the project, the production method and the size of the project (IEA 2010). The following table indicates the typical costs of new Canadian oil sands projects in 2010. Most oil-sand projects become economically interesting with oil prices at around 40–50 Euro/bbl. *In-situ* relates to deeper deposits (75 m and below) and *SAGD* is the acronym for Steam Assisted Gravity Drainage, which uses steam to facilitate the extraction of oil.

	Capital cost Euro/bl/d capacity	Operating cost Euro/bl	Economic price Euro/bl
Mining	32,000–46,000	16–23	33–52
In-situ primary	6,600	3–7	16–33
In-situ SAGD	20,000–26,000	13–20	30–53

From an environmental perspective, oil-sand production releases higher carbon dioxide emissions compared to conventional oil production due mostly to the large amount of energy needed during the production process. In numbers, between 175 and 300 kg of CO_2 is released per barrel of crude oil and between three and four barrel of water is needed to produce one barrel of crude oil, of which 95 % can be reused (IEA 2013b). In terms of land area, 140,000 km^2 of northern and eastern Alberta were exploited in 2010 to extract oil from oil-sands, which is larger than the combined area of Switzerland and Austria together.

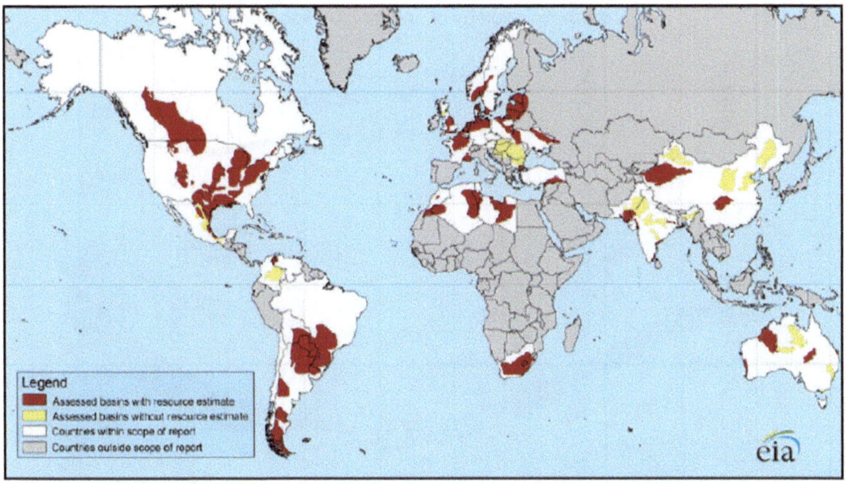

Fig. 2.42 World shale gas resources (from US Energy Information Administration)

2.7.3 Shale Gas

Shale gas is natural gas trapped in shale formations of very low permeability. Therefore, it can not be produced by the same technology as conventional gas, e.g. from gas caps. However, shale gas is very abundant on most continents, see Fig. 2.42, and has become an increasingly important source of natural gas especially in the United States where around 30 % of the producable gas is assumed to be shale gas. Shale gas exploration and production is today one of the fastest growing trends within the petroleum industry worldwide, and shale gas has been found in regions with no previous hydrocarbon production.

Shale gas is a so called dry gas which consists primarily of methane, with small amounts of higher hydrocarbon gas components plus nitrogen and hydrogen sulfide. Shale is a fine-grained, clastic sedimentary rock formed from mud that is a mix of flakes of clay minerals and tiny fragments (silt-sized particles) of other minerals, especially quartz and calcite. Shale gas is the result of organic deposits being present during compaction. The shale rock acts both as source rock and reservoir rock, contrary to conventional gas reservoirs where a migration of the hydrocarbons from the source rock to a reservoir rock takes place. A schematic illustration of both processes is shown in Fig. 2.43.

Production of shale gas started in the 1920s in Ohio, USA using vertical drilled wells. Shale has low matrix permeability, so gas production in commercial quantities

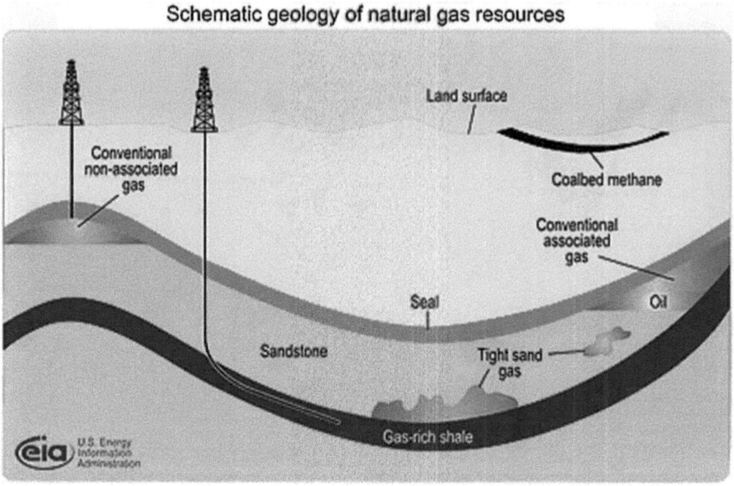

Fig. 2.43 Schematic geology of natural gas deposits (from US Energy Information Administration)

requires fractures to provide permeability. The natural fractures are predominantly vertical, so a large number of wells had to be drilled to drain a reservoir. For many years gas production from shales were considered unprofitable. But in the 1980s two new technologies emerged which revolutionized shale gas production—*horizontal drilling* and *hydraulic fracturing*. Horizontal wells reaching as far as 3,000 m create a maximum borehole surface area in contact with the shale. This makes it possible to produce below cultivated land and even underneath cities. A few vertical wells may be the starting point for a large number of horizontal production wells.

Hydraulic fracturing (fracking) is the main difference when producing gas from shale and from a conventional gas reservoir. A liquid at high hydraulic pressure is pumped into the formation which results in fracturing of rock, usually in the direction normal to the least stress in the formation. At large depth, the cracks have a tendency to become vertical and the influence of the hydraulic cracking may propagate several hundred meters, see Fig. 2.44. The fracturing strategy can be tailored for each specific shale formation by numerical simulations based on the special geometry and lithology of the petroleum bearing strata. This opens for large volumes of shale gas to flow to the wellbore and to be produced to the surface.

In an *unconstrained* world, the economics of shale gas is attractive. It currently costs between 2 to 6 Euro to produce one MMBtu of shale gas. This compares well to the production cost of one MMBtu of traditional gas, which amounts to between 0.1 and 6 Euro (IEA 2013b). The word *unconstrained* in the sentence above is important as most of the shale gas is so far produced in the United States and in Canada. With attractive economics, shale gas has been developed rapidly, especially in the US and the supply in natural gas increased drastically. In 2013, the US do not have the possibility to export its natural gas and prices have started to diverge from the price

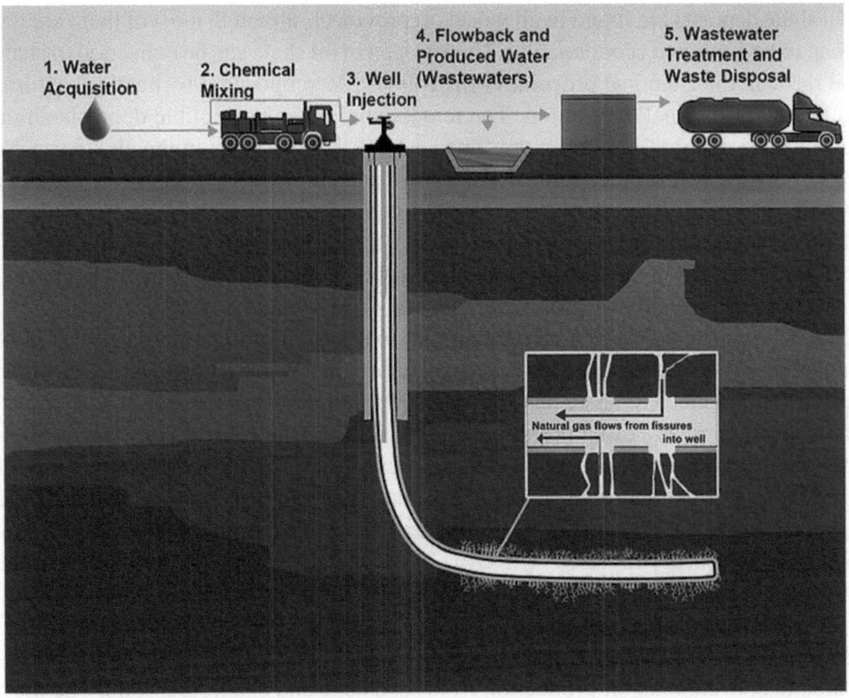

Fig. 2.44 Schematic depiction of hydraulic fracturing for shale gas; from U.S. Environmental Protection Agenecy (EPA)

of natural gas in other markets (see Fig. 2.34), limiting the margins of shale gas producers in the US.

Shale gas reserves are large, at least 210 trillion cubic meters, or enough to satisfy the world current demand for natural gas for another 60 years.

Hydraulic fracturing has raised environmental concerns about ground water contamination, risks to air quality, migration of gases and hydraulic fracturing chemicals leaking to the surface, mishandling of waste, and the health effects of all these, as well as its contribution to raised atmospheric CO_2 levels by enabling the extraction of previously sequestered hydrocarbons. In a recent report (posted to the web in November 2010) by the U. S. Environmental Protection Agency (EPA) on emission factors for greenhouse gas emissions by the oil and gas industry, EPA concluded that shale gas emits larger amounts of methane, a potent greenhouse gas, than does conventional gas, but still far less than coal.

2.7.4 Shale Oil

Oil shale is an organic-rich fine-grained sedimentary rock containing significant amounts of *kerogen* (a solid mixture of organic chemical compounds) from which technology can extract liquid hydrocarbons (shale oil) and combustible oil shale gas.

Oil shale deposits are found in all world oil provinces, although most of them are too deep to be exploited economically. The kerogen in oil shale can be converted to shale oil through the chemical processes of pyrolysis (decomposition by heating), hydrogenation, or thermal dissolution. The temperature when perceptible decomposition of oil shale occurs depends on the time-scale of the pyrolysis; in the above ground retorting process the perceptible decomposition occurs at 300° C, but proceeds more rapidly and completely at higher temperatures. The ratio of shale gas to shale oil depends on the retorting temperature and as a rule increases with the rise in temperature. Modern in-situ process, which involves heating the oil shale underground, may take several months of heating, decomposition may be conducted as low as 250 °C. Such technologies can potentially extract more oil from a given area of land than ex-situ processes, since they can access the material at greater depths than surface mines can. Oil shale has also been burnt directly as a low-grade fuel.

Depending on the exact properties of oil shale and the exact processing technology, the retorting process may be water and energy extensive. A critical measure of the viability of extraction of shale oil lies in the ratio of the energy produced by the oil shale to the energy used in its mining and processing, a ratio known as "Energy Returned on Energy Invested" (EROEI). A 1984 study estimated the EROEI of the various known oil-shale deposits as varying between 0.7–13.3. Global technically recoverable oil shale reserves are substantial with the largest reserves in the United States, Russia and Brazil. However, the production of shale oil is very limited compared to conventional oil production from sandstone and chalk reservoirs. The most important producers are Estonia, Brazil, China, and to some extent Germany and Russia. Oil shale gains attention as a potential abundant source of oil whenever the price of crude oil rises. At the same time, oil-shale mining and processing raise a number of environmental concerns, such as land use, waste disposal, water use, waste-water management, greenhouse-gas emissions and air pollution. The reserves of oil shale are large, estimated to about 4.8 trillion barrels. The economics of extracting this resource strongly limit its attractiveness for the time being.

2.7.5 Coal Conversion

Coal liquefaction is a general term referring to a family of processes for producing liquid fuels from coal. Specific liquefaction technologies generally fall into two categories: direct (DCL) and indirect liquefaction (ICL) processes. Indirect liquefaction processes generally involve gasification of coal to a mixture of carbon monoxide and hydrogen (syngas) and then using the so-called Fischer-Tropsch process to convert the syngas mixture into liquid hydrocarbons. By contrast, direct liquefaction processes convert coal into liquids directly, without the intermediate step of gasification, by breaking down its organic structure with application of solvents or catalysts in a high pressure and temperature environment. Since liquid hydrocarbons generally have a higher hydrogen-carbon molar ratio than coals, either hydrogenation or carbon-rejection processes must be employed in both ICL and DCL technologies.

As coal liquefaction generally is a high-temperature/high-pressure process, it requires a significant energy consumption and large capital investments. Thus, coal liquefaction is only economically viable at historically high oil prices, and therefore presents a high investment risk.

Most coal liquefaction processes are associated with significant CO_2 emissions, resulting either from the gasification process or from generation of heat and electricity that serve as energy inputs to the reactors. High water consumption in water-gas shift or methane steam reforming reactions is another adverse environmental effect. On the other hand, synthetic fuels produced by coal liquefaction processes tend to be 'cleaner' than naturally occurring crudes, as e.g. sulfur compounds are not synthesized or are excluded from the final product.

2.7.6 Methane Hydrates

Methane is a natural gas (CH_4) which exists in solid form together with frozen water, a so-called *methane hydrate*. The water molecules form, under certain conditions, cages which contain methane molecules. This results in large caps of methane hydrates which are deposited and stable on the sea floor at sufficient depths and in the ground. The map in Fig. 2.45 shows known locations of methane caps on the sea floor. The locations are closely correlated with the border zones of the continental plates. Natural deposits of methane hydrates also exist on Earth in colder regions, such as permafrost areas. There has been a long-term commercial production of natural gas from methane hydrates in Siberia.

The first enterprise to excavate methane hydrate from the seabed was stared up in March 2013 about 70 km off the Atsumi Peninsula, in central Japan. A drill ship from Japan Oil, Gas, and Metals National Corporation (JOGMEC) will drill 1,000 m into a 300 m thick layer of methane hydrates. The next step will involve inserting a large pipe down into the well in order to separate methane hydrate into methane gas and water. This is not straight forward, but theoretical analysis as well as laboratory experiments, has shown that CO_2 is able to stabilize the hydrate better than methane does, Graue et al. (2006) and Ersland and Graue (2010). Thus, if CO_2 is pumped into the hydrate deposits, methane will be released and CO_2 may be captured. This is of course a very interesting process since it has the potential of solving two problems: hopefully a commercial production of methane and storage of CO_2 in safe formations. To test this out US Department of Energy (DOE) in 2012, in conjunction with the oil company ConocoPhillips and JOGMEC will pump CO_2 down a well in Prudhoe Bay, Alaska, into a hydrate deposit. If all goes as planned, the CO_2 molecules will exchange with the methane in the hydrates, leaving the water crystals intact and freeing the methane to flow up the well. Conventional wells in the Prudhoe Bay gas fields contain a very high concentration of carbon dioxide, about 12 % of the gas, some of it is being pumped back into the conventional wells as a drive gas to maintain pressure.

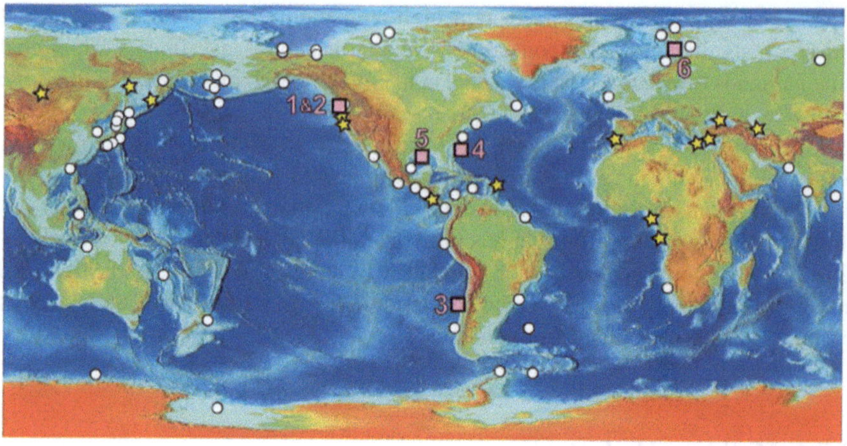

Fig. 2.45 Global distribution of known offshore gas hydrate reservoirs (Pohlman et al. 2009)

Conservative estimates imply that the energy potential of methane hydrates is huge—at least twice as much as the existing amounts of conventional oil. But commercialization is still a long way off. The United States has no urgent need to mine methane hydrates because of the increasing success of shale gas production. Japan, on the other hand, has few other fossil-fuel resources and plan to start longer tests of methane production from the Nankai Trough in 2015.

Fascinating as it is, mining methane is a risky business. Many geologists suspect that gas hydrates play an important role in stabilizing the seafloor. Drilling in these oceanic deposits could destabilize the seabed, causing vast swaths of sediment to slide down the continental slope. Hydrates also tend to form along the lower margins of continental slopes, where the seabed falls away from the relatively shallow shelf toward the abyss. The roughly sloping seafloor makes it difficult to run pipeline. But perhaps the biggest concern is how methane hydrate mining could affect global warming. Hydrate deposits naturally release small amounts of methane. Once methane is in the atmosphere, it becomes a greenhouse gas 20 times more efficient than carbon dioxide at trapping solar radiation. Some experts fear that when drilling in hydrate deposits it will be problematic to keep the released methane under control. The gas works itself skyward, either bubbling up through permafrost or ocean water, until it is released into the atmosphere. This could cause catastrophic releases of methane that would greatly accelerate global warming. There may also be a dangerous feedback effect. As the global temperature increases, more methane will be released from the artic tundra, which raises the temperature further.

With these very large reservoirs of unconventional fossil energy in the ground, it is far from certain that the "oil age" will end with the depletion of conventional oil resources. It is quite certain that the resource base is large enough to extend the "oil age" for up to another 100 years. As for development of other energy sources, the use of these non conventional energy sources will become a matter of price vs.

environmental impact. At the time we reach the world peak of conventional oil, it seems likely that the unconventional reserves will be one option which will be considered against nuclear and or renewable alternatives. In the following chapters, we will describe these alternatives.

2.8 Exercises

1. **Coal fired power plant**

A coal fired power plant has 1 GW electricity production when working at full capacity at a thermal efficiency of 35 %. The coal has a heating value of 30 MJ per kg and a sulfur (S) content of 2 % (by weight), a nitrogen (N) content of 1 % by weight and a mineral content (ash) of 10 %. Assume the coal consists of carbon with atomic mass 12 u, sulphur has mass 32.066 u, nitrogen 14.006 u, oxygen (O) has weight 15.994 u. Note: 1 u is equal to 1.66×10^{-27} kg.

(a) How much electricity can the plant produce per year?
(b) How much coal does the power plant consume each year?
(c) How much CO_2, SO_2 and NO_2 does the plant emit per year?
(d) Chose a realistic electricity price (in Norway, eg. NOK 0.5 per kWh) and calculate the value of 1 year output with full production.
Assume in a "zero interest rate country" that a device to capture and store CO_2 (CCS) from the plant will cost GNOK 5 (= 5 billion kr) in capital costs to install. This device requires 25 % of the total energy produced for operation. Assume further that operation of the plant (with or without) CCS costs 10 % of the value of the electricity production (in average) and that the initial cost of the plant is 10 % of the price of the electricity produced in a 20 year period.
(e) What is the total cost of the plant in a 20 year perspective with and without CCS?
(f) Discuss necessary conditions for plant owners to install the CCS technology? Write down "pros and cons".

2. **Developing an oil reservoir**

A small land based and essential horizontal oil reservoir has an area of approximately $A = 0.76\,\text{km}^2$, according to a geological model. The average thickness of the oil bearing zone is $h = 35$ m. Retrieving sandstone core plugs by test drilling shows that the porosity is $\phi = 23$ % and further core analysis indicate that the oil saturation is $S_o = 0.72$. There is no indication of a gas cap on top of the oil, but gas may be dissolved in the oil. When brought to the surface under standard conditions, dissolved gas will be released resulting in a shrinkage of the oil volume compared to the oil volume in the pressurized and temperature elevated reservoir. The *oil formation factor* B_o is the ratio between the oil volume in the reservoir and the oil volume and the stock tank oil volume at the surface. For this particular reservoir $B_o = 1.09\ \mathrm{Rm}^3/\mathrm{Sm}^3$.

(a) Calculate the Stock Tank Oil Originally In Place (STOOIP) for this reservoir in Sm^3 (Standard cubic meters).
(b) Assuming an oil price of 92 USD/bl, what is the gross value of the oil in the reservoir? (1 barrel = 159 l)
The production strategy is to produce the oil by a simple water drive, i.e. pumping water into the reservoir through 4 injection wells and producing the oil in a production well centred between the injection wells.
(c) Will it be possible to produce all the oil in the reservoir by this method? Which physical factors are limiting the recovery?
The investment costs for drilling and well completion will be 7×10^5 USD and the operational and maintenance costs are assumed to be 7.5 USD/bl. The reservoir engineers expect the recovery rate without any extra well stimulation to be 35 % and the production is started at a rate of 70 bl/day.
(d) The oil company is uncertain about the future oil price. Make a plot of the annual profit from this field as a function of time based on an oil price of 85 USD/bl. Assume the capital costs to be paid back in 5 years at an interest rate of 7 %.
After two years of production the investors decide that the return on their investment is unsatisfactory and ask the oil company to consider enhanced oil recovery (EOR) by injecting surfactants in the water drive.
(e) Discuss how this can improve the oil recovery.
(f) The cost of the surfactant is 20 USD/kg and its concentration in the injected water is planned to be 0.5 %. The oil production can then be increased to 100 bl/day without water breakthrough in the production well. What must the oil price be to make this programme profitable?

3. **Fuel cost natural gas**

Assume a natural gas price of 2.7 Euro/mmBtu.

(a) Calculate the expected fuel cost per MWh of electricity produced for that plant, assuming a thermal efficiency of 55 %, a discount rate of 5 %, an escalation rate of 1 % and plant life T = 30 years.
(b) Now, building a 250 MW plant would require an investment of Euro 125 millions. Assuming an economic plant life of 30 years, a capacity factor of 85 % and a discount rate of 5 %, calculate the LCOE of that plant. Operation and maintenance costs would amount to Euro 56 millions for the first year of operation, increasing at a 1 % rate annually.
(c) Taking the LCOE of question (b) as the expected electricity price you will receive over the next 30 years, you fear that the introduction of intermittent renewable energy generating capacity will harm the profitability of your plant. Such technologies have a marginal cost close to 0 Euro/MWh, which means that in hours with strong wind and sun, your plant will not be compet-

itive (electricity price < fuel + variable operation and maintenance costs of your natural gas-fired power plant). Taking this into account, you estimate that your capacity factor will decrease to 60 %. What is the LCOE of that plant with this new capacity factor.

(d) Using your answer in (c) can you tell why gas-fired plant owner see the deployement of wind and solar as a threat?

4. **Fuel cost coal**

Assume a coal import price of 70 Euro/metric ton, which is the same regardless of which coal type you import. Expecting that coal prices will be stable over the next 40 years, calculate the fuel cost per MWh of electricity produced for a plant with a thermal efficiency of 37 % if:

(a) the coal used is lignite (use Table 2.1 to make an assumption on the heat content of lignite).
(b) the coal used in anthracite (use Table 2.1 to make an assumption on the heat content of anthracite).
(c) if using a fuel with higher heat content leads to better economics, can you think of an element which would push coal-fueled power plants to use fuel with lower heat contents?

References

Aleklett, K., Qvennerstedt, O., Lardelli, M.: Peeking at Peak Oil. Springer, Berlin (2012) (doi:10.1007/978-1-4614-3424-5_2)

Blas, J.: Coal prices flare as australian floods cut supply. FT.com, 11 Jan 2011

BP.: Statistical Review of World Energy 2013. http://www.bp.com/content/dam/bp/pdf/statistical-review/statistical_review_of_world_energy_2013.pdf, 2013

Cassedy, E.S., Grossman, P.Z.: Introduction to Energy—Resources, Technology and Society, 2nd edn. Cambridge University Press, Cambridge (1998)

Clo, A.: Oil Economics and Policy. Kluwer Academic Publishers Group, Dordrecht (2000)

Northwest Power Planning Council.: Natural gas combined-cycle gas turbine power plants, Aug 2002

Deutsch, C.V., McLennan, J.A.: Guide to SAGD (Steam Assisted Gravity Drainage) Reservoir Characterization Using Geostatistics. Centre for Computational Geostatistics (2005)

Ersland, G., Graue, A.: Natural gas. In: Natural gas hydrates, 16p. SCIYO, Rijeka, Croatia (2010)

Gibbins, J., Chalmers, H.: Carbon capture and storage. Energ. Policy **36**(12), 4317–4322 (2008). doi:10.1016/j.enpol.2008.09.058

Graue, A., Kvamme, B., Baldwin, B.A., Stevens, J., Howard, J., Aspenes, E., Ersland, G., Huseb, J., Zornes D.R.: Environmentally friendly CO_2 storage in hydrate reservoirs benefits from associated natural gas production. SPE, OTC 18087, SPE, : Offshore Technology Conference (OTC) in Houston. OTC paper 18087, 2006

Helseth, L.E.: Fysikk og norsk oljeproduksjon. Fysikkens Verden **3**, 81–84 (2012)

Hubbert, M.K.: Nuclear energy and the fossil fuels. In: Spring meeting of the Southern District, San Antonio, Texas, Mar 1956

IEA.: Natural Gas Market Review. OECD Publishing, Paris (2009) (doi:10.1787/nat_gas_rev-2009-en)

IEA.: World Energy Outlook 2010. OECD Publishing, Paris (2010) (doi:10.1787/weo-2010-en)
IEA.: Coal Information. OECD Publishing, Paris (2011a) (doi:10.1787/coal-2011-en)
IEA.: Electricity Information. OECD Publishing, Paris (2011b) (doi:10.1787/electricity-2011-en)
IEA.: World Energy Outlook 2011. OECD Publishing, Paris (2011c) (doi:10.1787/weo-2011-en)
IEA.: Coal Information. OECD Publishing, Paris (2012) (doi:10.1787/coal-2012-en)
IEA.: World Oil Statistics. OECD Publishing, Paris (2013a) (doi:10.1787/data-00474-en)
IEA.: Resources to Reserves 2013—Oil, Gas and Coal Technologies for the Energy Markets of the Future. OECD Publishing, Paris (2013b) (doi:10.1787/9789264090705-en)
Jensen, J.T.: LNG versus pipelines—economic and technical pros and cons. In: Presentation to the Far North Oil & Gas Forum, Calgary, Nov 2007
Lindner, E.: Chemie für Ingenieure, p. 258, 11th edn. Lindner Verlag, Karlsruhe (1997)
NPD.: Well-filled toolbox for improved oil recovery, 2009
NPD.: Petroleum resources on the norwegian continental shelf, 2013
OECD.: Natural Gas Information. OECD Publishing, Paris (2011) (doi:10.1787/nat_gas-2011-en)
OECD.: Natural Gas Information 2013. OECD Publishing, Paris (2013) (doi:10.1787/nat_gas-2013-en)
OECD/Nuclear Energy Agency.: Projected costs of generating electricity (2010) (doi:10.1787/9789264084315-en)
Pohlman, J.W., Bauer, J.E., Canuel, E.A., Grabowski, K.S., Knies, D.L., Mitchell, C.S., Whiticar, M.J., Coffin, R.B.: Methane sources in gas hydrate-bearing cold seeps: evidence from radiocarbon and stable isotopes. Mar. Chem. **115**(1–2), 102–109 (2009). doi:10.1016/j.marchem.2009.07.001
WEC.: World Energy Resources: Survey. World Energy Council, London (2013). ISBN:978-0-946121-29-8
World Bank.: World Bank commodity price data (Pink sheet). Washington, DC, Aug 2013. http://econ.worldbank.org/WBSITE/EXTERNAL/EXTDEC/EXTDECPROSPECTS/0,,contentMDK:21574907menuPK:7859231pagePK:64165401piPK:64165026theSitePK:476883,00.html
Zerpa, L.E., Queipoa, N.V., Pintosa, S., Salager, J.-L.: An optimization methodology of alkaline surfactant polymer flooding processes using field scale numerical simulation and multiple surrogates. J. Petrol. Sci. Eng. **47**, 197–208 (2005)

Chapter 3
Nuclear Energy Systems

Abstract This chapter describes the basic processes of present and potential future nuclear power generation technologies. A brief outline of the historical development of nuclear power in the last century is first presented, followed by a short introduction to the nuclear physics needed to understand the fission process and the working principle of nuclear reactors. Later, the costs related to investments, operation, resources, storage and decommisioning of reactors are examined. Finally, a discussion on possible new nuclear technologies is provided at the end of this chapter.

Keywords Fission · Fusion · New nuclear technologies

3.1 Historical Development

The systematic study of the atomic nucleus started out in Europe and in the USA in the early 1930s. As World War II approached, physicists developed an understanding of the composition of the nuclei and of the basic nuclear processes. It was then realised that nuclear transformations, in certain cases can release an order of a million times more energy per atom than chemical reactions. Consequently, physicists on the German and on the Allied side understood that nuclear reactions can be utilised in bombs with a similar order of magnitude in the energy released by nuclear bombs compared to conventional bombs based on chemical reactions. The first nuclear reactor was developed under strict wartime secrecy and was demonstrated to work in Chicago, see Fig. 3.1 in 1942. The report to Washington DC was cryptically transmitted as "The Italian navigator has landed in the New World and found the natives very friendly". Here the "Italian Navigator" refers to the Italian leader of the project, Enrico Fermi. This achievement laid the ground for the Manhattan project which ended with the release of two atomic bombs over Hiroshima and Nagasaki in 1945 and lead to the sudden end of the war. After the war, development of nuclear reactors sparked off in parallel with the cold war and lead to today's arsenal of atomic bombs controlled by nearly ten countries.

P. A. Narbel et al., *Energy Technologies and Economics*,
DOI: 10.1007/978-3-319-08225-7_3,

Fig. 3.1 *Left* Drawing of the first nuclear reactor, University of Chicago 1942 (*Credit* M. A. Miller, Argonne National Laboratory) *Right* The Gundremmingen Nuclear Power Plant in Germany (*Photo* Felix König)

The first working nuclear reactors generating electricity for the grid appeared before 1960 and are today known as first generation prototype reactors. These reactors typically delivered 250 MW of capacity. In the 1960/1970s, a second generation of commercial reactors became available. These reactors typically delivered 1 GW of electricity. Most of the approximatively 440 nuclear reactors in the world currently active belong to this category. Reactors being built presently belong to a third generation of nuclear reactors with further improved security systems and technology.

A growing skepticism against nuclear power lead to an almost stop in the evolution of nuclear reactor technology for more than 20 years. The origin of the skepticism is major nuclear accidents in 1979 and 1986. The first one occurred at Three Mile Island (USA) and lead to small amounts of radioactive material into the environment. The second one took place in Chernobyl (Ukraine) and is the largest nuclear accident so far. It lead to a complete meltdown of the reactor core and significant emission of toxic radioactive materials.

In this century, a growing fear of global warming and the need for energy has lead to a renaissance for nuclear power. This has happened in spite of the recent earthquake and tsunami which destroyed a nuclear plant in Fukushima (Japan) in March 2011. This has lead to the extension of the economic life of several nuclear power plants originally planned for decommissioning and even political U-turns in countries where nuclear power were meant to phase out.[1]

Today the growth of nuclear power is significant in China and in India were almost 50 power plants are planned or under construction, and where intense nuclear research programmes, for example regarding new fuel, are ongoing. The reactors currently under construction can be denoted generation III since they have improved inbuilt safety systems, better design and waste control compared to generation II reactors.

[1] A referendum in Sweden in 1980 resulted in a parlamental decision to close down all reactors before 2010. Only one reactor, at Barsebäck, has been shut down. The remaining reactors covers today about 45 % of Swedens electricity consumption.

An interesting example in this context is the Finnish project at Olkiluoto. Here an advanced new nuclear reactor is beeing built which will be the first Generation III+ reactor constructed in the world. This reactor is planned to start operating around 2016. It started almost 10 years ago, and to get an idea of the magnitude of the project it requires more than 4,000 employees and is the largest industrial project in Northern Europe. A nuclear waste repository in stable rock structures 400 meters below the surface is under construction and will be connected to this reactor.

The status of nuclear power in the future is often called "uncertain" since the "nuclear *angst* (or fear)" in the public has shown to have a decisive influence when alternatives (mostly cheap fossil fuels) are available. The following factors will likely be important regarding the role of nuclear fission in the future:

- Energy supply alternatives. Both Germany and Sweden have had strong public opinions towards closing down their nuclear ractors. The lack of alternatives keep (most of) the nuclear plants running.
- Availability of fuel, waste handling and reprocessing facilities.
- Price: It is always difficult to chose the expensive alternative. So far, nuclear energy has proven both stable, competitive with fossil fuels and cheaper than renewables such as wind and solar energy.
- Future nuclear accidents. The recent accident in Japan resulted in a growing concern. A new serious nuclear accident will very likely make nuclear power more unwanted.
- New nuclear "safe" technologies. Development in nuclear power production may lead to future safe technologies with a drastic reduction in the waste and its life time. When/if such technologies will appear, and if they appear at an acceptable price per kWh generated, it may lead to a complete public turnaround regarding nuclear energy.

3.2 Basic Nuclear Physics

The building blocks of matter were briefly described in Chap. 1, Sect. 1.4 and illustrated in Fig. 1.4. Recall that the atomic nucleus is built up of *nucleons*, which are either positively charged *protons* or neutral particles called *neutrons*. They have approximately the same mass, 1.67×10^{-27} kg, the neutron being slightly heavier than the proton. An element's position in the periodic table is defined by the number Z protons in the nucleus, which is the same as the number of negative electrons circulating it. The number N neutrons in the nucleus may vary for each element resulting in different *isotopes* for that element. The total number of nucleons is $A = Z + N$. A given nucleus, called a *nuclide*, is then uniquely defined by A, Z, N and is denoted by the element name and the number of nucleons A, e.g. Carbon-12 ($^{12}_{6}C$). The naturally occurring elements on Earth range from Hydrogen ($Z = 1$) to Uranium ($Z = 92$). The different elements have varying number of isotopes, from 1 to 10 ($_{50}Sn$). Isotopes may be either stable or *radioactive*. A few natural occurring isotopes are radioactive, e.g. $^{238}_{92}U$, but most known radioactive isotopes are either the result of nuclear reactions or fission.

A radioactive isotope has some excess energy, which means it will undergo change over time. It disintegrates or decays with a characteristic time called the *half-life* $T_{1/2}$. During this process Z and N may change and energy is radiated as gamma radiation (γ) or particles, such as α-particles (^4_2He), β-particles (positive or negative electrons) or in some cases nucleons. If we have a sample of originally n_0 radioactive nuclei at time $t = 0$, the number that will disintegrate per time unit will be proportional to the number of nuclei n that has yet not decayed, hence:

$$\frac{dn}{dt} = -\lambda n \tag{3.1}$$

where λ is called the *disintegration constant*. This can easily be integrated to yield:

$$n = n_0 \exp(-\lambda t) \tag{3.2}$$

The number of radioactive nuclei falls off exponentially and the half-life is the time it takes until $n = n_0/2$. This leads to $T_{1/2} = \ln 2/\lambda$. Half-lives are different for each specific radioactive isotope and range from microseconds to billions of years. We define *activity*, or the decay rate R, as the number of disintegrations per time unit:

$$R = -\frac{dn}{dt} = \lambda n_0 \exp(-\lambda t) \tag{3.3}$$

From this we see that the activity also falls off exponentially and it is inversely proportional to the half-life, hence a short half-life implies a strong radioactivity. We measure the activity in the unit *becquerel* (Bq) which is one integration per second. This is a small unit since even a small sample of radioactivity will contain a huge number of disintegrating nuclei. The old unit used for activity is the *curie* (Ci) which is a large unit; 1 Ci = 3.7×10^{10} Bq.

If a human is exposed to nuclear radiation, some or all of the radiation energy may be absorbed in the body's tissue. We define the *absorbed dose D* as the energy absorbed per mass unit of matter, which is measured in the unit *gray* (Gy) = 1 J/kg. Traditionally the unit *rad* has been much used; 1 rad = 0.01 Gy. When the radiation hits an atom one of its electrons may be emitted, which changes the atom's ability to enter into bindings with other atoms to form molecules. This again influences the cell reproduction, and serious damage to the tissue, such as cancer, may result. The biological effects vary by the type and energy of the radiation and the organism and tissues involved. We define the *equivalent radiation dose* $H = QD$, where the quality factor Q depends on the type of radiation, ranging from $Q = 1$ for X-rays, γ-rays, and β-rays to $Q = 20$ for α-particles and neutrons. The equivalent radiation dose is measured by the unit *sievert* (Sv) which is J/kg. The recommended maximum equivalent radiation dose per capita for the population as a whole is 1 mSv/year.

A whole-body exposure to 5 Gy of high-energy radiation at one time usually leads to death within 14 days. However, in cancer radiation therapy a radiation dose of 60–80 Gy directed exactly to the tumor, may be given, fractionated over several weeks.

3.3 Nuclear Fission

The force between the protons in the nucleus, the so-called *Coulomb force*, is repulsive. But the nucleons are bound together by another force, *the nuclear force,* which prevents them from flying apart. The *total binding energy* E_b of a given nuclide would be the energy needed to separate the nucleus into its components—the protons and neutrons. It can readily be calculated from the difference in mass between Z protons plus N neutrons and the mass of M_A of the nucleus:

$$E_b = (Zm_p + Nm_n - M_A)c^2 \tag{3.4}$$

which is based on Einstein's famous mass—energy formula $E = mc^2$; c being the speed of light. The total energy of the bound system (the nucleus) is therefore less than the combined energy of the separated nucleons, and the energy difference was released when the nucleus was formed.

The *binding energy per nucleon* E_b/A varies with the number of nucleons or mass number A as shown in Fig. 3.2. We observe that the curve has its maximum around $A = 60$. This implies that when two light nuclei collide and transform into a heavier element, the mass of the heavy element is less than the mass of the two separate elements before the reaction. The excess mass, is mostly released as kinetic energy of the nucleus after the reaction—and this kinetic energy can in principle be harvested in a reactor. This process is called *fusion* and the process can continue between more massive nuclei until we reach the nuclei with nucleon numbers around 60. On the heavy element side, the inverse process can take place. This process is called *fission*: A single heavy nucleus, like Uranium-238 ($^{238}_{92}$U) with 238 nucleons, can spontaneously split into two lighter elements with the release of energy, most of it as kinetic energy on the two fission fragments. This is the working principle of today's nuclear fission reactors. Fusion, on the other hand, is the process for energy production in the Sun.

Spontaneous fission with observable half-lives occurs in the heavy isotopes like $^{232}_{90}$Th, $^{235}_{92}$U and $^{238}_{92}$U, but these nuclei preferably disintegrate by $\alpha-$decay or $\beta-$decay instead. However, *induced fission* can be initiated by the capture of a *thermal neutron*, which is a neutron with kinetic energy of 0.025 eV ($\simeq 4.0 \times 10^{-21}$ J) or less. The neutron will excite a collective vibrational mode in the nucleus so it becomes elongated and finally breaks in two. The two resulting nuclei X and Y are called *fission fragments* and the nuclear reaction may be written:

$$^{1}\text{n} + {}^{235}_{92}\text{U} \rightarrow {}^{236}_{92}\text{U}^* \rightarrow \text{X} + \text{Y} + \text{neutrons} \tag{3.5}$$

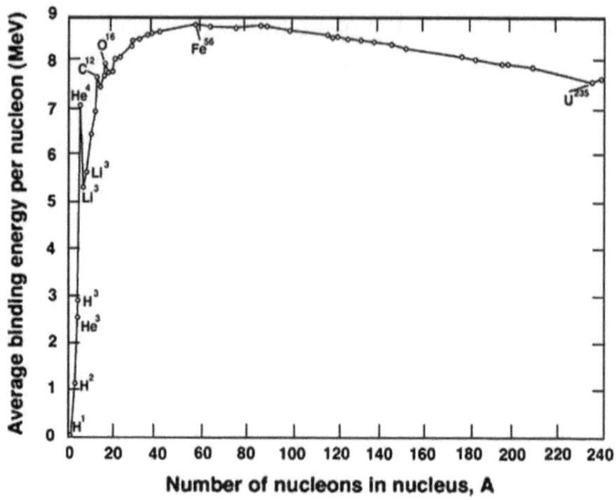

Fig. 3.2 Binding energy per nucleon shown as function of nucleon number A

where $^{236}U^*$ is an intermediate excited state, called a compound nucleus, which lasts only for about 10^{-12} s before it splits into the two fission fragments and a number of neutrons, usually 2 or 3. There are many channels into which $^{235}_{92}U$ may fission. A typical reaction is:

$$^{1}n + ^{235}_{92}U \rightarrow ^{141}_{56}Ba + ^{92}_{36}Kr + 3\ ^{1}n \tag{3.6}$$

The average number of neutrons in such reactions is about 2.5. As we shall see in the next section, it is these neutrons which may sustain the chain reaction in a nuclear reactor or in atomic bombs. When $^{235}_{92}U$ fissions more than 90 different fission fragments X and Y may be the result, always in pairs, one being somewhat heavier than the other. Figure 3.3 shows the distribution of the two fission fragments as a function of mass for three commonly fissile heavy nuclei. The fragments have a neutron excess immediately after the fission. Some neutrons "evaporate" instantaneously, but the fragments are still highly radioactive and decay to more stable nuclei in a succession of $\beta-$decays, emitting gamma radiations in the process.

From Fig. 3.3 we see that the fission fragment distribution is asymmetric with one peak around $A = 95$ and one around $A = 140$. These two regions contain the isotopes Strontium-90 ($^{90}_{38}Sr$) with a half-life of 29.1 years and Cesium-137 ($^{137}_{55}Cs$) with a half-life of 30 years. These two radioactive isotopes, among others, represent the problematic waste that come from all nuclear reactors, both because of their abundance in the fission yield curve, and their moderate, but not neglible half-lives. Hence the radioactivity from these two isotopes will be large for many decenniums.

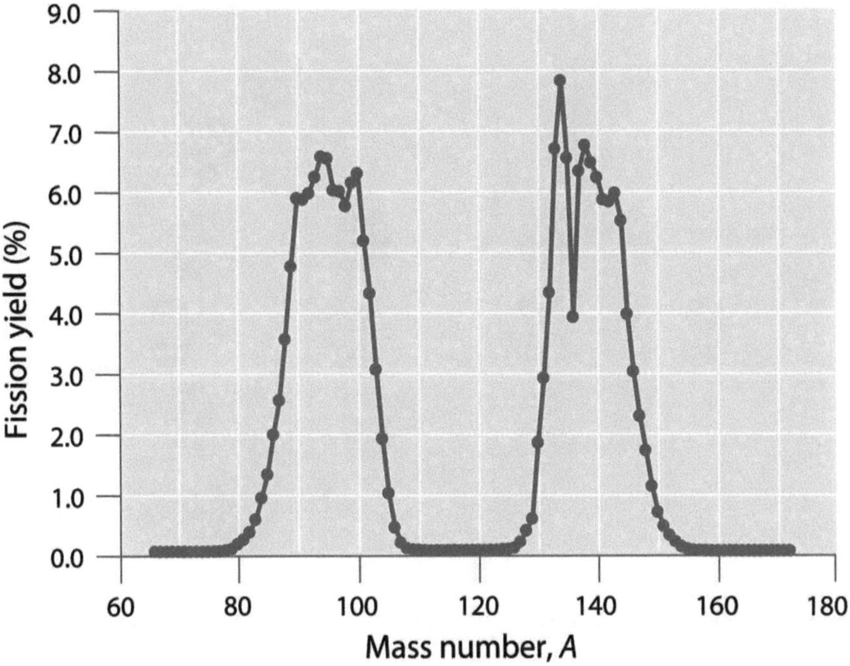

Fig. 3.3 Fission product yield as a function of mass A for fission initiated by thermal neutrons in ^{235}U

3.4 Nuclear Reactors

A nuclear reactor is a system designed to maintain what is called a *self-sustained chain reaction*. An average of one neutron emitted in each ^{235}U fission must be captured by another ^{235}U nucleus and cause that nucleus to undergo fission. This is illustrated in Fig. 3.4. As mentioned in the previous section, the neutron energy must be thermal for the ^{235}U isotope to be *fissile*. At this low energy the neutron spends much time in the vicinity of the nucleus, enhancing the probability for fission to take place. The original high energy neutron therefore must be moderated, i.e. slowed down. This is achieved by surrounding the fuel by a *moderator* (water or graphite) in which the neutron can undergo multiple collisions, slowing it down in each step. The other Uranium isotope, ^{238}U, is fissionable, but only for neutron energies above 1 MeV. Therefore, relatively few neutron induced fissions of ^{238}U takes place in a nuclear reactor and there will not be enough high energy neutrons to sustain a chain reaction in ^{238}U. Instead ^{238}U may capture a neutron to form ^{239}U, which eventually decays into Plutonium-239 ($^{239}_{84}$ Pu). This isotope may be used as bomb material.

A useful parameter for describing the level of reactor operation is the *neutron reproduction constant K* which is defined as the average number of neutrons from each generation of fission events in Fig. 3.4 that cause a new fission in the next

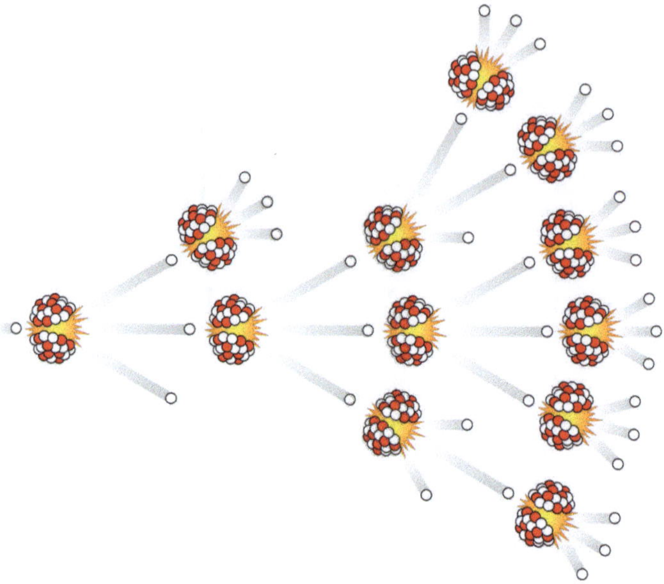

Fig. 3.4 Illustration of the chain reaction starting from a single ^{235}U nucleus and a neutron (white sphere)

generation. A self-sustained chain reaction is achieved with $K = 1$. The reactor is then said to be *critical*. When $K < 1$, the reactor is *sub-critical* and the chain reaction will die out. As mentioned above, an average of $\nu = 2.5$ neutrons are released in the fission of ^{235}U. Unless the number of neutrons in the reactor is controlled, the reactor may become *supercritical* and a runaway situation will occur. The number of neutrons in the reactor must therefore be controlled. There are several factors which reduce the number of available neutrons. They can be described by the so-called *four factor formula*:

$$K = \eta \times \epsilon \times p \times f \tag{3.7}$$

- some of the ν neutrons are captured without resulting in fission. η is the number which may become thermal neutrons ($\eta = 2.068$ for ^{235}U).
- some fast neutrons will cause fission in ^{238}U which are always present in the reactor fuel. The neutrons from this fission will multiply slightly. For natural Uranium $\epsilon \simeq 1.03$.
- p is the fraction of fission neutrons that manage to slow down from fission to thermal energies without being absorbed. A typical value of p is about 0.9.
- some neutrons may also be absorbed in the reactor components (moderator, cooling medium etc). f is the fraction that does not get absorbed here. A typical value of f is also about 0.9

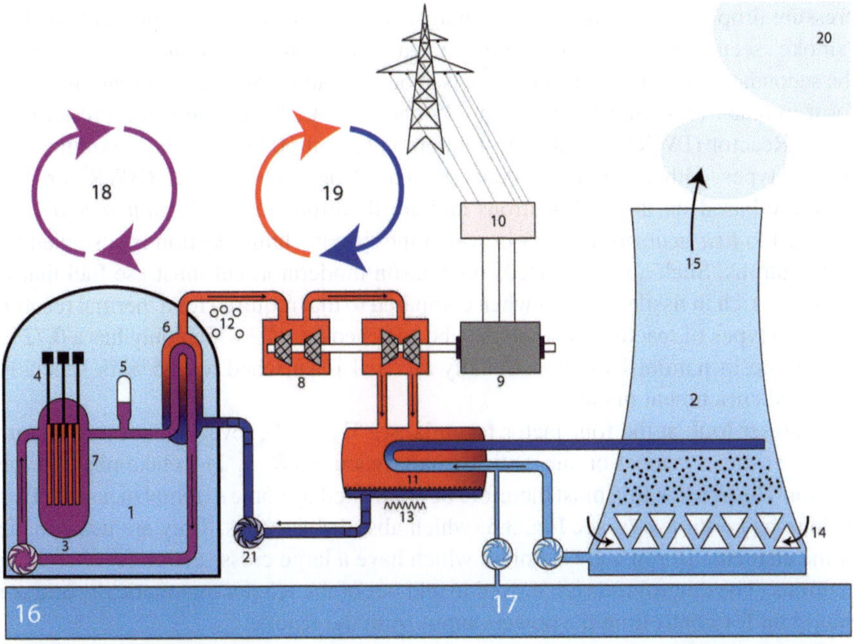

Fig. 3.5 Principle for electricity generation in a pressurized water reactor (PWR). The main numbers on the figure refer to: 1. reactor block, 2. cooling tower, 4. control rod, 7. fuel element, 8. turbine, 9. generator, 10. transformer, 11. condenser, 21. pump. Steffen Kuntoff, Wikimedia Commons

Each fission event releases about two hundred million electronvolts (200 MeV) of energy. By contrast, most chemical oxidation reactions such as burning coal or TNT release at most a few eV per event. So, nuclear fuel contains at least ten million times more usable energy per unit mass than does chemical fuel. The energy of nuclear fission is released as kinetic energy of the fission products and fragments, and as electromagnetic radiation in the form of gamma rays. In a nuclear reactor, the energy is converted to heat as the particles and gamma rays collide with the atoms that make up the reactor and its working fluid, usually water or occasionally heavy water. Figure 3.5 illustrates the working principle for a pressurized water reactor (PWP). The reactor core contains the fuel in the form of Uranium oxide pellets loaded in long rods inside the reactor pressure vessel. The vessel is filled with pressurized water which acts both as a coolant and as a moderator. The purpose of a moderator is to slow down the high energy fission neutrons to thermal neutrons, so that they can be captured by a new Uranium nucleus and initiate a new fission.

The water in the reactor vessel (the primary loop) heats a secondary closed loop water cycle. This water is then transformed into steam, which powers a turbine where the steam loses about half of its kinetic energy (pressure drops) and is further cooled to water in a condenser system. The turbine run an electric power generator, much in the same way as in a coal plant, compare Fig. 2.2. The condensation causes a maximum pressure drop across the turbine which again maximizes electricity production. The "smoke" seen from nuclear reactor plants is a result of the water condensation. Inside the secondary water loop, the pressure is about 150 atm which allows water in liquid form to reach temperatures between 275 and 315 °C, hence the name Pressurized Water Reactor (PWR). It is the most common (about 60 %) of today's commercial reactor types. Other coolant systems are based on boiling water (BWR) or gas. Both are based on thermal neutrons and are therefore called *thermal reactors*, as opposed to *fast neutron reactors* in which the fission chain reaction is sustained by fast neutrons. Such a reactor needs no neutron moderator, but must use fuel that is relatively rich in fissile material when compared to that required for a thermal reactor. For both types of reactors the fuel must be enriched in ^{235}U which only has a 0.72 % abundance in natural Uranium. Usually the fuel is enriched to 3–5 % in specially designed enrichment plants.

A closer look at the four factor formula, see Eq. (3.7), reveals that these factors do not ensure a critical or sub-critical chain reaction ($K \leqq 1$) to take place in the reactor. The neutron flux must therefore be controlled by some external measure. This is done by *control rods*, see Fig. 3.5, which absorbs neutrons. They are usual made of the elements Boron and Cadmium which have a large cross section for absorbing neutrons. The control rods are moved in and out of the reactor automatically and are important for controlling the power output from the reactor.

Due to the large amount of energy released per nuclear reaction, the amount of fuel needed per year to run a nuclear reaction is very small. To produce 1 GW · day, 2.7×10^{24} nuclei must fission. However, this corresponds to a ^{235}U mass of only 1.035 kg. While a 1 GW coal based power plant requires several hundred thousand tonnes of coal per year for the fuel process, the amount of ^{235}U needed in a nuclear plant is less than a tonne. For comparison, the waste of a coal based power plant in terms of greenhouse gas (GHG) is about 3 times the mass of the fuel which amounts to millions of tonnes per year, while in a nuclear plant it remains on the order of a single tonne waste per year with no GHG released in the electricity production. But on the downside, the present nuclear waste remains radioactive for 100,000 to a million years.[2] We will discuss this issue separately in Sect. 3.6.

3.5 Present Use, Resource Considerations and Forecast

If most clean energy technologies are being built at an increasing pace worldwide, nuclear power is (still) the exception. Rapid growth in the nuclear capacity installed occurred until 1986 before slowing down after the famous Chernobyl accident. In fact, the total operational capacity increased by less than 10 % between 1995 and 2012 and the leading countries in terms of installed capacity are the United States

[2] Increased antropogenic CO_2 emissions will remain in the atmosphere for a few thousand years before it is reabsorbed by the ocean and living plants.

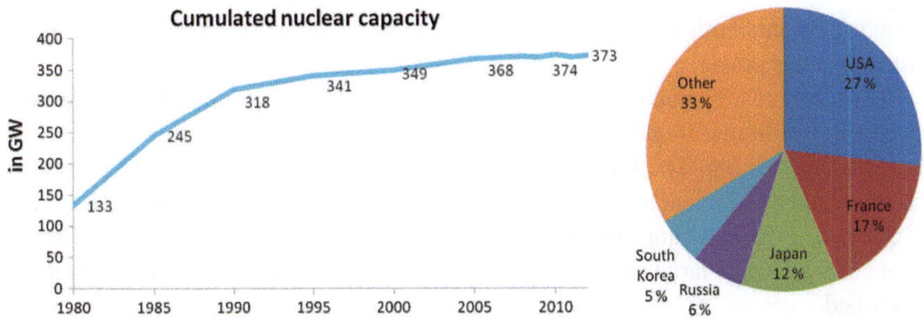

Fig. 3.6 Installed nuclear capacity until 2012 and leading countries in terms of nuclear capacity at the end of 2012 (IAEA 2013)

with over 100 GW of installed capacity at the end of 2012, France (63 GW), Japan (44 GW), Russia (24 GW) and South Korea (21 GW), see Fig. 3.6.

Note that the installed nuclear capacity in Japan at the end of 2012 amounted to 44.2 GW, altough the 50 reactors in Japan "only" produced 17.2 TWh of electricity.[3] This low level of electricity production is the result that only few (around 2) reactors were running in 2012. The other reactors are waiting for permission to restart after they have been reinforced in order to comply with new safety regulations following the Fukushima accident. This is likely to take another few years.

If the operational capacity is stagnating overall, local trends differ widely. China has multiplied its nuclear capacity five folds in the last fifteen years (13 GW in 2012), while India (4 GW) and South Korea (20 GW) doubled theirs. Over the same time period, the operational nuclear capacity in France and in the United States has barely increased and even decreased in Bulgaria, Japan, Canada, Germany and in the United Kingdom.

Compared to other clean technologies, only large hydropower projects at present can match or exceed the size of a nuclear power plant. The smallest nuclear plant installed to date is located in Russia and has a net capacity of 44 MW. This plant was built between 1974 and 1976. All the subsequently built nuclear power plants have had a net capacity exceeding several hundreds of MW, with the biggest being the Kashiwazaki-Kariwa nuclear power plant in Japan. The first reactor of the Kashiwazaki-Kariwa plant started generating electricity in 1985 and the last two of the seven units generated their first kWh in 1996. This power plant has a capacity of nearly 8 GW.

Recent technological progress has resulted in increased plant availability, higher capacity factors and improved efficiency. This progress has pulled down the levelized cost of energy generated from nuclear power plants. As a matter of fact, the nuclear generating capacity has increased by one percent annually over the past two decades, whereas the total electricity generated increased by 2–3 % per annum. On the other hand, stricter safety requirements are pulling capital costs up. Given the limited

[3] Assuming a normal capacity factor for nuclear reactors of 85 %, Japan should have generated around 330 TWh of electricity from its 50 reactors.

number of plants built in the past years, it is difficult to identify a clear trend in the cost of nuclear power.

Fuel resources for nuclear fission are limited but the exact quantities are largely uncertain. Uranium is found in the Earth's crust usually as uranium oxide (U_3O_8), and harvested by mining. The known economically recoverable resources at present is estimated to be sufficient to cover our present needs for another 100 years. However, ^{235}U exists in substantial amounts in sea water. The resource base may thus be orders of magnitude larger if techniques for uranium enrichment from the sea can be developed.

The future of nuclear power is highly uncertain. Some studies suggest that the global nuclear energy capacity may double by 2030 which would be equivalent to building up to another 400 GW of new generating capacity. Other studies plan on a decrease in the global installed nuclear capacity over the same time period. Recent developments, especially the nuclear accident in Fukushima in March 2011 may deter future nuclear installations. However, due to the cost-competitiveness of this technology, the arrival of more efficient and more secure nuclear technologies, decreasing fossil-fuel reserves, the potential of nuclear power in mitigating climate change and an increase in electricity demand from developing countries, it is likely that the global installed nuclear capacity will expand in the future.

New technology options, to be discussed at the end of this chapter, may also provide options for new nuclear power production, since these may completely shift the public opinion towards nuclear power.

3.6 Cost

3.6.1 Capital Costs

The most common PWR technology has been chosen to discuss the costs of fission nuclear power as it is the most widely spread nuclear technology around the globe. All cost estimates given hereafter refer to this technology.

Capital costs for a typical PWR reactor ranges between 2,500 and 4,000 Euro/kW of installed capacity. These costs account for a high share (between 65 and 77 %) of the LCOE. Nuclear power is therefore a capital intensive technology. For nuclear reactors, the engineering and construction costs are the main components of the capital costs. Regulations can potentially also be an important component of the capital costs. Strict safety regulations require additional performing safety systems to be installed. In some plants, *seven* independent active and passive safety systems have to be installed to reduce the risk of a chain reaction accident to a level close to zero.

The plant size plays a role as economies of scale are significant in the nuclear industry. For this reason, larger units often achieve comparatively lower capital costs per unit of installed capacity.

In the past, disputes about licensing or oppositions from local communities and environment preserving associations (if not properly built, nuclear plants may, e.g.

affect the water source they use for cooling purposes) or changing regulations due to events such as Fukushima, have delayed the construction and completion of nuclear power plants, thus resulting in an increase in capital costs for several plants.

Based on these elements, several studies differentiate the discount factor used for capital cost compared to other costs, since the risk related to the construction of the plant is comparatively higher compared to other risks. In the present chapter, no such differentiation takes place.

3.6.2 Operation and Maintainance Costs

Operation and maintenance costs (O&M) account for 11–25 % of the levelized cost of electricity and they increase over time as the plant gets older. The O&M costs are influenced particularly by changing regulatory requirements. After the Fukushima accident in 2011, strenghtened requirements have been put in place to ensure that operational nuclear power plants can resist devastating natural events, thus leading to an increase in O&M costs. Minimal requirements for in-service inspections are also being set by regulatory requirements.

3.6.3 Resource Costs

The main primary material for nuclear power plants is uranium. Ruled by the laws of supply and demand, the uranium price has experienced fluctuations in the past. Between 1988 and 2004, the uranium price remained under 80 Euro/kg in real prices. Between 2005 and 2007, the prices have been multiplied by nearly a factor of ten due to several factors including increasing demand from China and India and expected supply/demand gap in the future, the flood of the largest underdeveloped high grade deposit in the world and climbing prices of oil and natural gas. Since 2008, uranium prices have decreased to around 100–200 Euro/kg, with a recent decreasing trend due, among other factors, to a lower than expected demand for Uranium in Japan. The evolution of the uranium price between 1983 and 2013 is illustrated in Fig. 3.7.

It has to be noted that this price corresponds to a spot price and only represents about 20 % of the exchanges in the market. Most contracts are long term, covering periods from three to fifteen years. Fluctuations in the price of uranium do not significantly impact the final price per kWh generated as fuel costs only account for 7–14 % of the levelized cost of electricity and as only a third of the fuel cost is the price of uranium oxide, the rest being for the enrichment and fuel fabrication. Therefore, nuclear power is, to some extent, insensitive to changes in uranium prices, which is an advantage over the fossil fuel technologies for which fuel costs account for a higher share of the final costs.

On the demand side, the general trend is characterized by an increasing demand for uranium although the Fukushima accident led to a temporary reduction in demand.

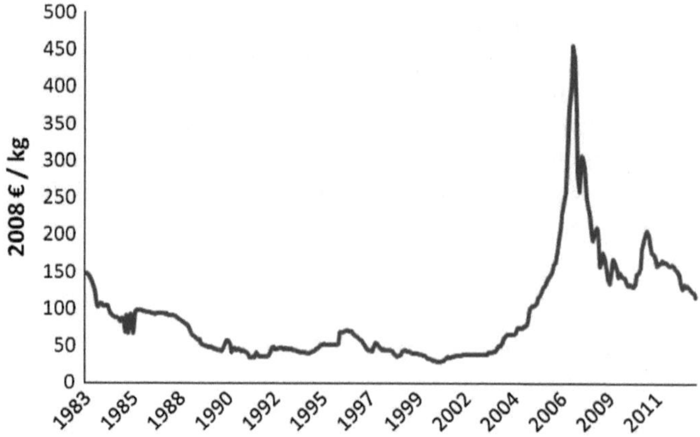

Fig. 3.7 Historical price in Euro^{2008} per kg of uranium, August 1983–July 2013 index mundi (2013)

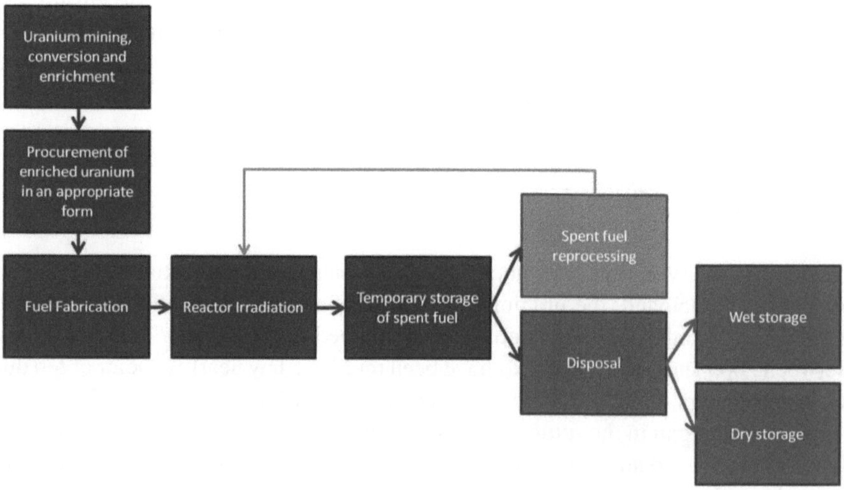

Fig. 3.8 Nuclear fuel cycle IAEA (2009)

The production of uranium oxide covers 78 % of the utilities annual requirements (63,000 tonnes in 2010), the balance being from secondary sources such as stockpiled uranium, ex-military weapons-grade uranium and plutonium[4] and recycled uranium and plutonium from spent fuel.

On the supply side, the general trend includes a steady increase in production, especially after the 2007 peak price when many facilities were reopened. The main countries supplying uranium oxide are Kazakhstan (covering 33 % of the world's demand), Canada (18 %) and Australia (11 %).

[4] Weapons-grade material can be diluted with a factor 25:1 given the high quality of the material.

3.6.4 Storage Costs

Figure 3.8 illustrates the nuclear fuel cycle. It ends with its disposal and storage (red boxes).

It is estimated that 290,000 tonnes of heavy metals (or spent fuel) had been discharged by 2006 by the nuclear reactors around the world. Of this amount, roughly a third has been reprocessed and the remaining had and still has to be stored under strict conditions to ensure long term safety. This quantity is growing and could reach 455,000 tonnes of heavy metal by 2020.

These heavy metals can either be stored on site or in away-from-reactor storage facilities. Two technologies are available for storing heavy metals; the wet (in pools) and dry technologies which can easily be distinguished by the cooling medium used. Storage costs of nuclear waste are dependent on a number of factors. The type of nuclear waste stored will impact the costs as will the storage technology chosen, the size of the facility and its degree of modularity. A higher degree of modularity leads to a significant economic advantage over time. Dry technologies have a higher degree of modularity over wet technologies and can therefore more easily keep track with the spent fuel quantities.

Three categories of costs are attached to storage: capital costs including the design engineering and construction costs, O&M costs including spent fuel loading, monitoring and spent fuel unloading and the decontamination and decommissioning costs (D&D) at the end of the useful life of the facility. The latter comprehends all costs related to the restauration of the site to its original condition.

The World Nuclear Association reported in early 2011 that the storage costs account for up to 10 % of the cost per kWh of electricity generated by the nuclear power plant. In other words, the storage costs of spent fuel amounts to about 0.5–0.8 Eurocents/kWh generated.

3.6.5 Decommissioning Costs

The reader will probably have read in newspaper articles that if appropriate decommissioning costs of a nuclear plant were included in the price per kWh, the electricity generated in a nuclear power plant would certainly not be as economically attractive as it seems to be today. Under the LCOE method, decommissioning costs can easily be integrated and the next paragraph will help the reader understand that even though these costs can be enormous in absolute terms, they barely impact the cost per kWh generated from nuclear power if discounted using the traditional approach.

Let us assume that the decommissioning of a nuclear power plant costs as much as the capital costs to build it. Thus, a 1,600 MW power plant with a capital cost of 2,800 Euro/kW will require Euro 4.48 billions for its decommissioning, which appears to be humongous. Deciding whether to invest or not, an investor will discount

Table 3.1 Levelized cost of electricity for nuclear PWR plants

Label	Nuclear PWR
LCOE Euro/MWh	55–81
Overnight cost %	68–79
O&M costs %	11–25
Fuel costs %	7–14

this amount. With a discount rate of 10 % and an economic plant life of 60 years, the present value of decommissioning this nuclear plant amounts to:

$$4.48 \text{ GEuro} \cdot \frac{1}{(1.1)^{60}} = 14.7 \text{ MEuro} \tag{3.8}$$

which is much smaller. This number translates to around 9 Euro per kW of installed capacity. Hence, decommissioning costs are too far in time to be a decisive criterion for an investor. However, these costs will occur and a way of including them into the LCOE is to input a decommissioning cost equivalent to a share of the capital costs of the power plant. Along with other reports (IEA/NEA 2010), we assume a decommissing cost equal to 15 % of the capital costs of nuclear power in this book.

3.6.6 Electricity Generation Costs of Nuclear Fission

Capital cost (including decommissioning costs), O&M costs and fuels costs (including storage costs) are the major components of the electricity generation cost of nuclear fission. As the capital costs account for a big share of the electricity generation costs, it is important to ensure a high degree of plant availability and a high capacity factor, since the levelized costs of nuclear power will decrease if the total quantity of electricity generated increases.

The length of the economic plant life of a nuclear power plant will impact significantly the LCOE. Existing studies use very different plant lives, from twenty years to 60 years. Since most nuclear power plants have been in operation for over 30–40 years, an economic plant life of 60 years is used here.

Assuming a high degree of availability and a capacity factor of 85 %, the levelized cost of electricity from PWR nuclear reactors amounts to 5.5–8.1 Eurocents/kWh. Nuclear power is thus cost-competitive in most electricity markets given the assumptions used to estimate its various costs (Table 3.1).

At a time when the threat of climate change is a concern for countless economies and when the perspective of putting a price on carbon is seen as a likely possibility, it is interesting to notice that such price on carbon will not affect the economics of nuclear power. As stated earlier in the chapter, nuclear power does not release GHG in the atmosphere during the generation process and a price on carbon will thus not result in an increased price of the levelized cost of nuclear power.

Another interesting fact about nuclear power is that most externalities (storage and treatment of nuclear waste) are included in the final price, which is

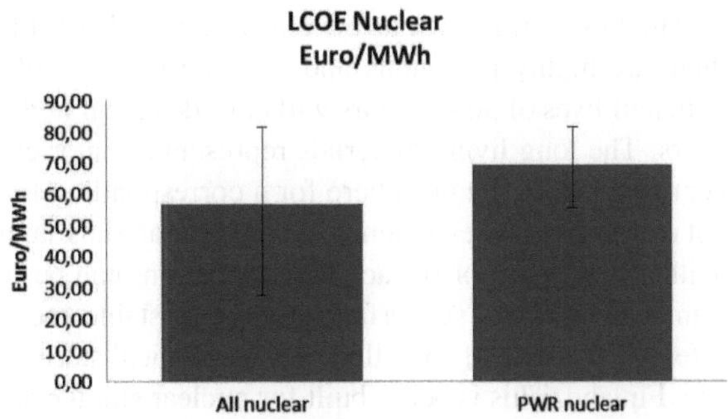

Fig. 3.9 Levelized costs of electricity from PWR nuclear powerplants

Table 3.2 Long lived isotopes and their half-lives

Isotope	Half-life	Relative fraction in waste (%)
^{85}Kr	10.8 year	0.2
^{90}Sr	29.1 year	4.5
^{137}Cs	30 year	30.2
^{99}Tc	0.21 Myear	6.2
^{93}Zr	1.5 Myear	5.5

clearly not the case for power generated from fossil fuels as the cost of cleaning the atmosphere for the damage caused by releasing carbon dioxide is typically not added to the cost of electricity generated from coal or natural gas.

3.7 Nuclear Waste; Safety and Storage

As mentioned in the previous section, nuclear reactors of today are driven based on keeping control of the internal neutron production. If the control is lost (as in Chernobyl in 1986) a nuclear accident with dramatic consequences for the afffected population may occur. So even if nuclear reactors are safe and grave accidents extremely unlikely, the consequences of a potential accident can be more serious than accidents in other types of power plants. A single accident (i.e. Chernobyl) has halted the construction of new nuclear reactors after 1980 despite the fact that nuclear energy is one of the safest branches of energy production in terms of fatal accidents per kWh produced. It is like flying, which is very safe with few accidents, but with dramatic accidents once in a while, the public opinion has turned away from this form of transport due to fear of accidents.

Another problem with present nuclear energy production is the waste. The fission products after the transformation of uranium is a mixture of quite toxic material with radioactive half-lives from a few days to millions of years. Those with the shortest half-life will decay while in the fuel inside the reactor, but many will still be active when the reactor has to be refueled, and will therefore appear as nuclear waste. Some of the longer lived isotopes are listed in Table 3.2.

The table shows just a few out of hundreds of different end products. Many of them are highly poisonous and must be kept out of the biosphere. The materials with half lives of 30–50 years will have desintegrated and vanished in a few hundred years. The long living materials represent a major challenge since they need to be kept away from the biosphere for a corresponding long time. Our civilisation does not really have any experience of storing materials safely for such time periods, since neither buildings nor surface based locations can be considered stable on such time frames. Therefore, subsurface storage in stable rock regions are considered most safe. An example of so-called deep geological storage sites are the "Onkalo" (cave) site, Finland. This is being built for nuclear storage about 400 meters below surface in a particularly stable granite rock, and it is expected to be in operation from around 2020 and will eventually contain sealed nuclear waste with potential for safe storage for millions of years. From this perspective it is sometimes claimed that this disposal is an acceptable solution to the nuclear waste problem.

Another option for nuclear waste is to transform it through nuclear transmutation. Here the waste material is bombarded by accelerated particles which collide with the material and transform it into nuclei which can be either handled in short term storage or used as new fuel. The economy of such processes is still uncertain, and research programs at present investigate how cost-effective nuclear transmutation can become effective. A final and more optimistic view is to consider the waste material as fuel for new future nuclear reactor technologies. For example, plutonium may undergo fission and produce energy when bombarded by fast neutrons. A reactor system being stable and using fast neutrons may thus represent the end of the nuclear waste problem. However, such reactors need to operate at much higher neutron energies than the thermal neutron reactors, and attempts to operate these reactors have not shown itself competitive so far.

A final, and perhaps the biggest concern with nuclear power is the possibility for proliferation of plutonium into nuclear bombs: Once a thermal reactor is available in a country, one also has a source for production of nuclear bomb material. Therefore the upbuilding of nuclear power in countries such as North Korea or Iran is considered a big threat by neighboring countries. Although nuclear power is constrained in order to limit this threat, because the basic processes are known, history has shown that countries which have the resources and ambitions to develop a nuclear arsenal can do it.

3.8 New Nuclear Technology

In many years to come, the thermal reactors will dominate the energy production from nuclear fission. However, it is clear from the preceeding sections, that present nuclear technology, with finite access to ^{235}U fuel, waste problems and possibility for plutonium proliferation, has disadvantages which makes it a less attractive power option for many countries. In this final section, we discuss a few alternative nuclear technology options which today exist only as plans or research projects. They have in common one aspect: If the technology can be proven, the resource potential for

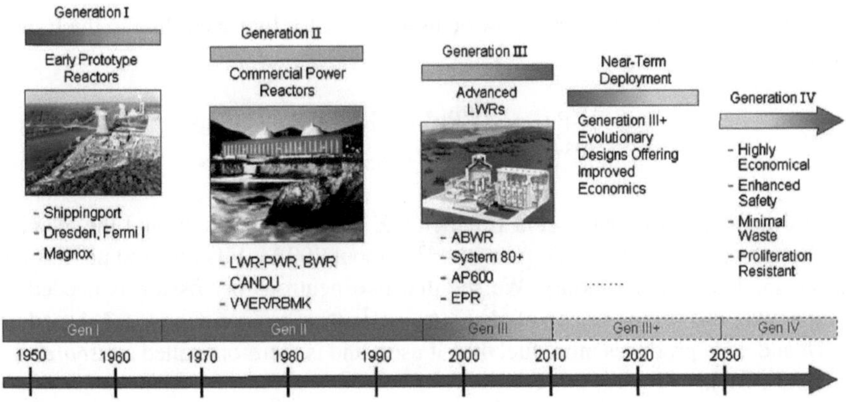

Fig. 3.10 Illustration of the different generations of nuclear reactors

energy production is in excess of thousands of years and in amounts which can cover the energy demand globally.

3.8.1 Fast Reactors—Generation IV

The "road map" through the different generations of nuclear reactors leading up to generation IV reactors is shown in Fig. 3.10. An international working group, Generation IV's International Forum (GIF) is a cooperative international endeavor organized to carry out the research and development needed to establish the feasibility and performance capabilities of the next generation nuclear energy systems. Apart from improved concepts and security systems of thermal reactors, the GIF group also envisage so called fast reactors. In such reactors, the fast neutrons from fission are used without moderation to undergo energy producing reactions using ^{238}U, ^{232}Th and even some of the waste from today's reactors. This will reduce the storage problem of nuclear waste from geological time scale to some hundred years, and will represent not only an end to the geological storage problem, but also offer the present nuclear waste a new life as future fuel. Furthermore, the burnup of plutonium in the reactor itself would set an end to the reactor as a potential "bomb material factory".

When ^{232}Th and ^{238}U are used as fuel, these two isotopes are first transformed to ^{233}U and ^{239}Pu respectively by neutron capture. The first step is the breeding of the two fissible isotopes

$$^{238}_{92}\mathrm{U} + n \rightarrow\ ^{239}_{92}\mathrm{U} \rightarrow\ ^{239}_{93}\mathrm{Np} + \beta^- + \overline{\nu} \rightarrow\ ^{239}_{94}\mathrm{Pu} + \beta^- + \overline{\nu}$$
$$^{232}_{90}\mathrm{Th} + n \rightarrow\ ^{233}_{90}\mathrm{Th} \rightarrow\ ^{233}_{91}\mathrm{Pa} + \beta^- + \overline{\nu} \rightarrow\ ^{233}_{92}\mathrm{U} + \beta^- + \overline{\nu}$$

where $\overline{\nu}$ is a chargeless and essential massless particle called an antineutrino. The two new isotopes have half-lives $T_{1/2} = 2.41 \times 10^4$ yr for ^{239}Pu and $T_{1/2} = 1.59 \times 10^5$ yr for ^{233}U, respectively. They can now be used as reactor fuel according to the fission reactions

$$^{1}\mathrm{n} + {}^{239}_{94}\mathrm{Pu} \rightarrow {}^{240}_{94}\mathrm{Pu}^{*} \rightarrow \mathrm{X} + Y + \text{neutrons}$$
$$^{1}\mathrm{n} + {}^{233}_{92}\mathrm{U} \rightarrow {}^{234}_{92}\mathrm{U}^{*} \rightarrow \mathrm{X} + Y + \text{neutrons}$$

Several hundred different fission fragments X and Y may result, and the average number of neutrons is $\nu = 2.3$. As with ^{235}U about 200 MeV is released per fission, but no moderator is necessary. We see that two neutrons per fission is needed as compared to one for the fission of ^{235} U. A fast neutron reactor with $\nu > 2$, based on ^{232}Th and ^{238}U produces more fuel than it uses, and is therefore called a *fast breeder reactor* (FBR).

The resource base of ^{232}Th and ^{238}U are of the order 100 times more common in the ground than ^{235}U. This imply that the resource base is so large that it can be considered as a renewable energy production source: It can cover the energy supply needed for a 10 billion Earth population for many thousands of years. Stable commercial reactors based on such principles are however unlikely to exist before 2030, and it is a long process of technology and system development needed in order to develop proven and safe concepts. Since what is needed is long term international research and technology projects, it seems unlikely that there will be fast progress as long as both nuclear generation III reactors are commercial competitive together with available cheap fossil based energy supply. However, there exist national and international development programs along these directions. Experimental fast breeder reactors have been built and operated in the United States, the United Kingdom, France, the former USSR, India and Japan.

An example of one of the generation IV concepts is the Gas-cooled fast reactor (GFR), shown in the left side of Fig. 3.11. The reactor fuel can be both uranium or thorium. The cooling system is planned to be a gas which can secure stable working conditions at high temperatures, 800 °C, for example He.

Several other cooling concepts exit for generation IV reactors, most notably the Sodium-cooled fast reactor (SFR) which uses the light metal sodium ($Z = 11$) in a liquid phase as coolant. One of the design challenges of an SFR is the risks of handling sodium, which reacts explosively if it comes into contact with water. However, the use of liquid metal instead of water as coolant allows the system to work at atmospheric pressure, reducing the risk of leakage. Still another fast reactor design which can work as a breeder is the Lead-cooled fast reactor (LFR) which is using melted lead as coolant with an outlet temperature of 550 °C or more. So far these reactor are merely theoretical designs and although smaller test reactors have been built, they can not be expected to produce energy on a commercial basis near 2030.

Another interesting example of potential future designs is "energy amplifiers" or accelerator driven reactors (ADS), which is a coupling between a reactor and a particle accelerator, see Fig. 3.12. The accelerator delivers fast protons to a material whose nuclei absorbs the protons, for example lead (which also acts as a coolant), and which emit a spectrum of neutrons from slow to fast. These neutrons drives nuclear

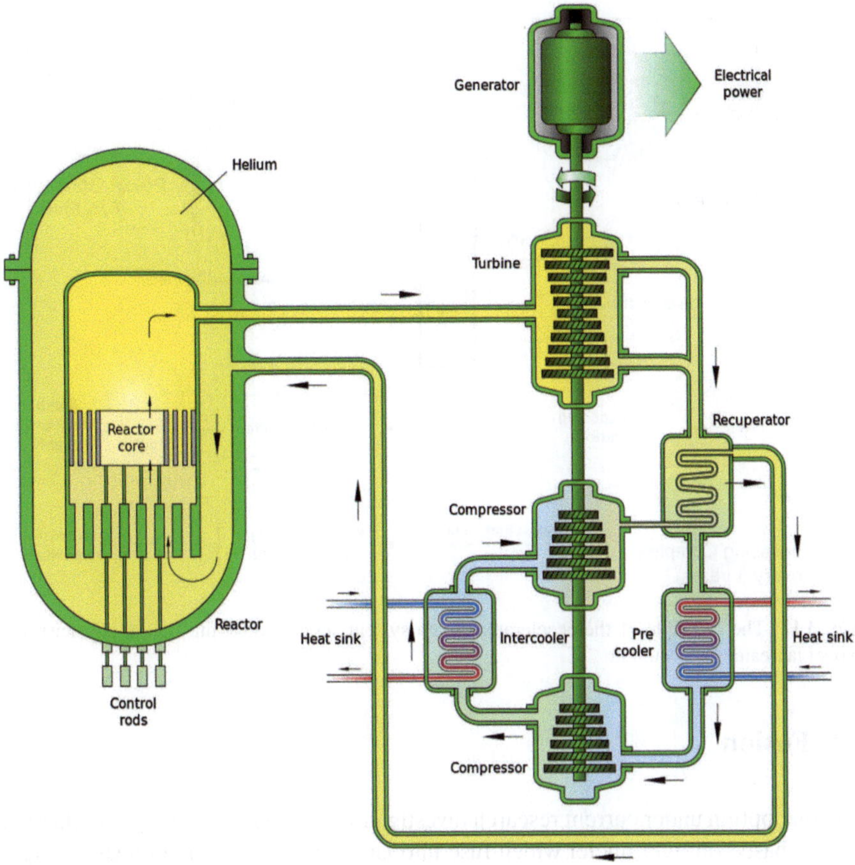

Fig. 3.11 A generation IV gas cooled fast reactor (GFR)

energy production in a thorium based fuel. Thorium is four times more abundant than uranium and no isotope separation of ^{235}U is needed for preparing fissile fuel. Both fast and slow neutrons lead to energy producing nuclear fission with minimal waste. In fact ,the energy amplifier will burn not only its own waste but could also burn waste from other reactors because the proton beam may transform the fission fragments to shorter-lived isotopes. A few percent of the energy production is used to drive the accelerator. Such reactor principles would be inherently safe since the reactor is subcritical: it needs the accelerator for energy production. Thus, simply by turning off the accelerator the nuclear reaction in the reactor will come to an immediate stop. An other advantage is that the proliferation problem is practically absent, since the system does not produce plutonium. A research project (MYRRHA) is going on in Belgium and is planned for full-scale operation in 2023–2024.

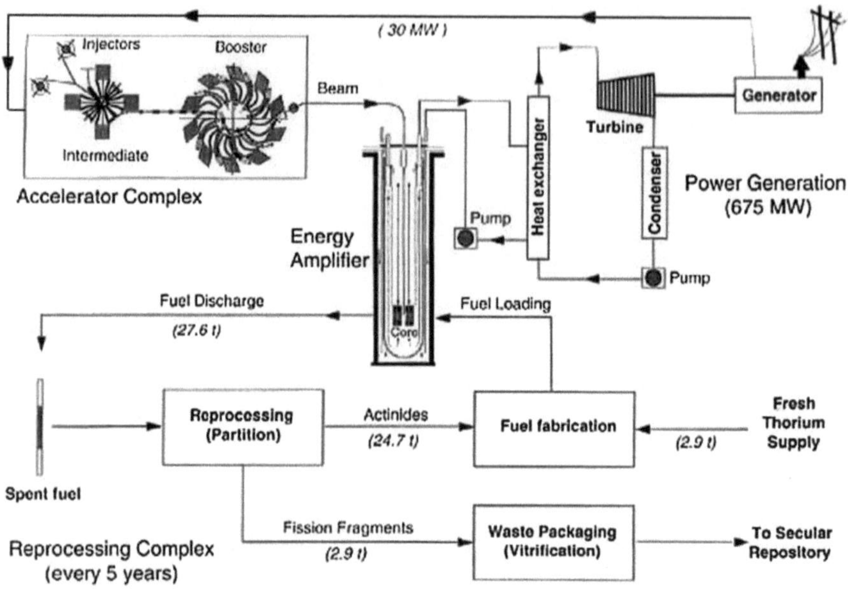

Fig. 3.12 The principle of the accelerator driven system suggested by the Italian physicist and Nobel laureate Carlo Rubbia

3.9 Fusion

Another option under current research investigation is *fusion*. The energy production is here based on light nuclei which fuse into larger compounds and release energy, cf. Fig. 3.2. The sun is a stable fusion energy "reactor" for our planet system. Helium is here produced from hydrogen at an almost constant rate in the following cycle of reactions

$$
\begin{aligned}
{}^{1}_{1}\mathrm{H} + {}^{1}_{1}\mathrm{H} &\rightarrow {}^{2}_{1}\mathrm{H} + e^{+} + \text{energy} \\
{}^{2}_{1}\mathrm{H} + {}^{1}_{1}\mathrm{H} &\rightarrow {}^{3}_{2}\mathrm{He} + \text{energy} \\
{}^{3}_{2}\mathrm{He} + {}^{3}_{2}\mathrm{He} &\rightarrow {}^{4}_{2}\mathrm{He} + {}^{1}_{1}\mathrm{H} + {}^{1}_{1}\mathrm{H} + \text{energy}
\end{aligned}
$$

Energy in the equations above refer to both radiation and kinetic energy of the final products. The release of freed energy per reaction cycle is about 20 MeV.

The task of building a small fusion reactor on Earth is faced with the problem of creating stable "sun like" conditions in a reactor. Temperatures on the order of millions of Kelvin are needed to give the light hydrogen atoms high enough kinetic energy to overcome the charge repulsion and allow the nuclei to come close together and react. There are two alternative reactions suggested for a fusion reactor: (D–D) reaction and a (D–T) reaction, where *deuterium* (D) is the hydrogen isotope with one neutron (${}^{2}_{1}\mathrm{H}$), and *tritium* (T) the hydrogen isotope with 2 neutrons (${}^{3}_{1}\mathrm{H}$). The

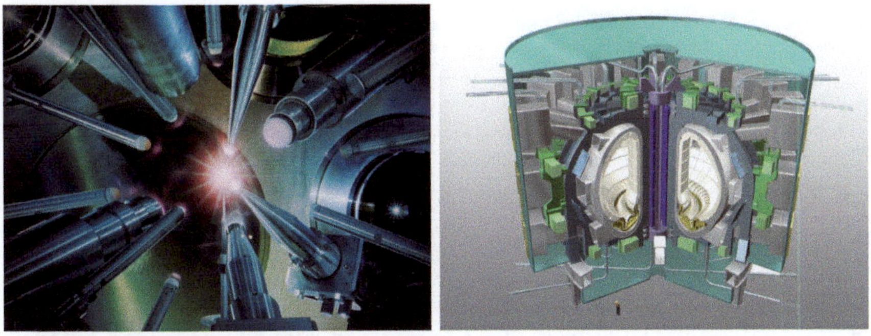

Fig. 3.13 The core of two possible future fusion reactors based on high power lasers (*left* credit Lawrence Livermore National Laboratory) and the Tokamak technology (*right*)

critical ignition temperature T_{ignit} is about 4×10^8 K equivalent to 35 keV for the (D–D) reaction and about 4.5×10^7 K equivalent to 4 keV for the (D–T) reaction.

In addition to the high temperature requirements, there are two other critical parameters that determine whether or not a thermonuclear reactor is successful: the ion density n and the confinement time τ, which is the length of time the ions are maintained at $T > T_{ignit}$. The *Lawson's criterion* states that a net energy output is possible under the following conditions:

$$n\tau \geq 10^{14} \text{ s/cm}^3 \text{ (D–T)}$$
$$n\tau \geq 10^{16} \text{ s/cm}^3 \text{ (D–D)}$$

In a *tokamak*, see Fig. 3.13, the idea is to keep a hot plasma (ionized light particles) inside strong magnetic fields with high enough plasma temperature to allow for fusion.

So far no such reactors have been able to produce more energy than it needs to heat and control the plasma. This is precisely the goal of the International Thermonuclear Experimental Reactor (ITER) project in France. Here the most advanced and powerful fusion reactor is under construction and planned to operate around 2020. The goal is to demonstrate that it is possible to produce energy in controlled forms from fusion. Specifically, the ITER reactor is designed for an energy output of 50 MW which is 10 times more energy than what is needed to operate it. The reactor fuel will be hydrogen isotopes which will be heated to 150 million Kelvin and meet the Lawson's criterion by keeping the plasma stable in the inner region of the tokamak by extremely strong magnetic fields. If the ITER projects succeeds it likely will lead to the "Next step" which will be the construction of a prototype machine for electricity production. Adding at least 20 years to this process it seems unlikely that tokamak fusion will be available for electricity production before 2050.

Another interesting development of fusion options is the *inertial confinement* system. In this setup very intense lasers are focused on a small pellet containing fusion fuel, for example water. When the lasers fire, the strong electromagnetic field

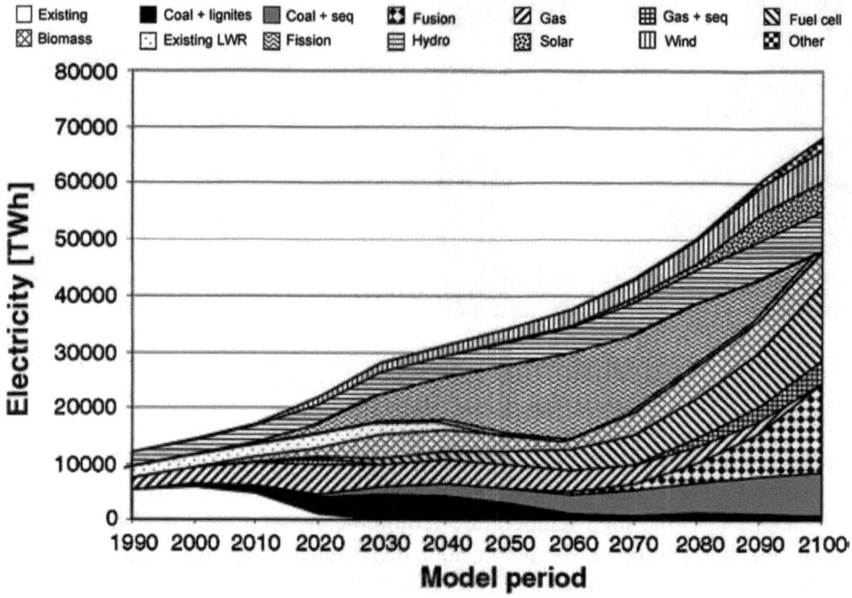

Fig. 3.14 Foresight of future electricity generation by fusion and other sources towards 2100. Lechon et al, Fusion Engineering and Design, 75, 1141, 2005

immediately tears the atoms and molecules of the pellet apart and creates a plasma. Then, a shock wave will implode and bring the protons of the fuel towards the centre of the pellet and close enough to allow for fusion processes. At the National Ignition Facility (NIF) in California, it is expected that this process will be demonstrated within a few years. If successful, the possibility for repeated laser firing and harvesting energy from this fusion concept will start, in competition with both tokamak concepts as well as new fission technologies. If either of these fusion technologies becomes "proven" and is developed into commercial reactors the power supply problem can almost be considered solved. The fuel (hydrogen) will cost almost nothing and the resource is infinite from a practical point of view and the technology comes almost without radioactive waste problems. To some extent fusion is a commercial threat to the present nuclear industry, but on the other hand the generation IV concepts will be a challenge for possible future fusion reactors with respect to price. In Fig. 3.14 these aspects are shown in a foresight for the global energy consumption. We here observe that fusion becomes a growing and important energy source from 2050, on the expense of both a phaseout of fossil energy and that generation IV reactors will not be built. Such foresights are generally extremely uncertain. What is clear is that most of the present nuclear reactors will have to close at some point. For safe and sufficient energy supply it is not unlikely that new nuclear technologies will replace the old ones, the open question is which technology will they be based on?

3.10 Exercises

1. **Nuclear decay**
The isotope Carbon-14, or ${}^{14}_{6}C$, is radioactive and has a half-life og 7730 years. If you start with a sample of 10^6 carbon-14 nuclei, how many will still be around after 22920 years?

2. **Nuclear activity**
The artificially produced radioactive isotope Iodine-131, or ${}^{131}_{53}I$, has a half-life of 8.04 days.
(a) Calculate the number of nuclei in 3.50 μg of this isotope at $t = 0$ (the time of production).
(b) What is the activity of this sample in Becquerels (Bq) at $t = 0$ and at $t = 24$ hours?

3. **Fission energy**
Calculate the fission energy of 1 kg natural Uranium. Remember that the fissile isotope is ${}^{235}_{92}U$ which has an abundancy of 0.72 % in natural Uranium. The disintegration energy per fission is 208 MeV. State your answer in MeV, Joules and kWh.

4. **Fusion energy**
About 1 of every 3,300 water molecules contains one deuterium atom.
(a) If all deuterium nuclei in 1 liter (L) of water fused in pairs according to the reaction ${}^{2}_{1}H + {}^{2}_{1}H \rightarrow {}^{3}_{2}He + n + 3.27$ MeV, how much energy in Joules (J) is liberated ?
(b) Burning gasoline produces about 3.40 J/L. Compare the energy obtainable from the fusion of the deuterium in a liter of water with the energy liberated from the burning of a liter of gasoline.

5. **Nuclear processes**
A slow neutron collides with ${}^{235}_{92}U$ and a nuclear reaction leads to the new nuclei ${}^{141}_{56}Ba$ og ${}^{92}_{36}Kr$. The mass of the particles are: n: 1.008665u, U:235.04393, Ba: 140.914411 u, Kr: 91.926156 u
(a) Set up the entire reaction and compute the released energy.
(b) The energy difference in the following two nuclear reactions are

$$^{235}\mathrm{U} + \mathrm{n} \rightarrow ^{144}\mathrm{Ba} + ^{89}\mathrm{Kr} + 3\mathrm{n} + 177\mathrm{MeV} \tag{3.9}$$

$$^{2}\mathrm{D} + ^{3}\mathrm{T} \rightarrow ^{4}\mathrm{He} + \mathrm{n} + 18.3\mathrm{MeV} \tag{3.10}$$

Estimate how many reactions per second is needed in a 1 GW power plant based on each of the reactions above when you assume a plant efficiency of 35 %.
(c) What is the mass of the spent fuel in a year? (Compare the number with the Coal plant, exercise 1, p. 110)
(d) Nuclear energy will increase their share of the global energy if fossil fuel becomes scarce/expensive, renewables cannot meet demand, global warming

causes supernational enforcement to drop/stop fossil plants and national goverments decides to be energy independent of the Middle East. Which reasons are most likely? Discuss each reason. Are there other reasons? What are possible reasons that will terminate nuclear power *forever*?
Constants:

$$c = 3 \times 10^8 \text{ m/s}$$
$$1\text{u} = 1.66 \times 10^{-27} \text{ kg}$$
$$1\text{eV} = 1.6 \times 10^{-19} \text{ J}$$

6. **Decommissioning of a nuclear reactor**
When dismanteling an old thermal nuclear reactor its core is found to contain radioactive materials 1 kg ^{85}Kr, 10 kg ^{90}Sr and 5 kg ^{99}Tc.
(a) How much of these materials is left after 10 years?
(b) How much is left after 1000 years?
(c) What is the activity after 1000 years?

7. **Economics of nuclear power**
The table below summarizes some key information about the economics of a new 'Advanced Gen III+' nuclear reactor.

Label	Unit	Quantity
Installed capacity	MW	1,350
Load factor	%	90
Cost to build the plant	Euro/kW	2,345
Fuel cost for the first year of operation	Euro/MWh	6.5
Levelized O&M costs for the first year	Euro	95,750,000
Discount rate	%	6
Escalation rate fuel cost	%	2
Economic plant life	years	60
Decommissioning costs[1]	Euro/kW	350

[1] The decommissioning costs are set equivalent to 15 % of the capital costs in accordance with the common practice, and can thus be inputed on top of the capital costs

(a) Calculate the LCOE of the nuclear power plant?
(b) Is any major cost, particular to nuclear power, missing from the table above?
(c) The policy maker wants to force solar power into the system. What will the likely impact be on the economics of nuclear power if wind power is deployed at a large scale?

References

IEA/NEA: Projected Costs Generating electr. (2010). doi:10.1787/9789264084315-en
IAEA. Costing of spent nuclear fuel storage (2009). ISBN 978-92-0-104109-8
IAEA. Nuclear power reactors in the world (2013). http://www-pub.iaea.org/books/IAEABooks/10593/Nuclear-Power-Reactors-in-the-World-2013-Edition
index mundi. Uranium monthly price–US dollars per pound (2013). http://www.indexmundi.com/commodities/?commodity=uranium

Chapter 4
Renewable Energy

Abstract This chapter describes the basic processes of the major renewable energy technologies. A brief outline of the historical development of renewable energy is first presented. Later, the working principles and the economics of wind energy, solar energy, hydro power, bioenergies will be described, followed by a review of other renewable technologies which have not reached maturity as of today. Finally, a discussion on the means available for carrying and storing energy is provided at the end of this chapter.

Keywords Solar power · Wind power · Hydropower · Biomass · Energy storage carriers

4.1 Historical Development

There has been a noteworthy change of pace in the rate of population growth before and after the invention of the steam engine. Before, the human population developed slowly, reaching 0.3 billion humans around 2000 years ago and passing the one billion mark at the start of the 19th century. The steam engine, driven by coal, became commercially available at that time and it opened up for faster transport, clearly outperforming horses and sail ships. In addition, it led to a fast change in human life, now based on industrialization. As seen in Fig. 4.1, the population has increased dramatically since then. The doubling from one to two billion took only a hundred years, while it took more than forty thousand years to reach the first billion. At the start of the 21st century we passed 7 billion (in 2011). Fortunately, the growth rate is now decreasing with an outlook for stabilization of the total population after 2050, although with significant uncertainty regarding the level at which the population will stabilize itself (United States Census Bureau 2008).

A more efficient use of fossil energy sources combined with a better technology has allowed the human population to grow to its present magnitude. To replace fossil

P. A. Narbel et al., *Energy Technologies and Economics*,
DOI: 10.1007/978-3-319-08225-7_4,

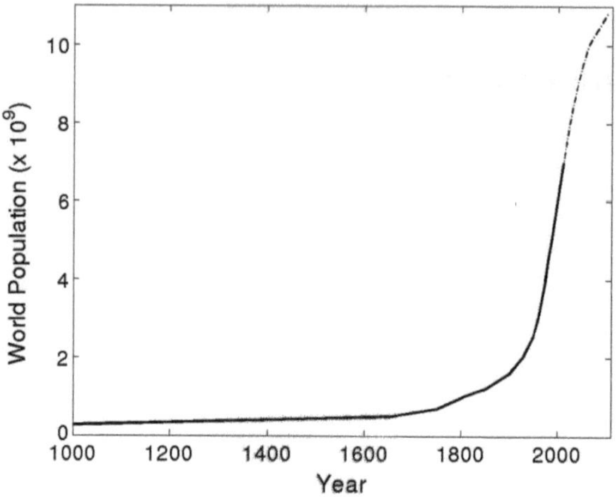

Fig. 4.1 Global human population development (*full line*) and a probable projection into the future (*dotted lines*)

fuels by alternative renewable and sustainable energy resources is truly a grand challenge, perhaps the greatest challenge in the history of humans ever.

Let us start with the definition of two commonly used words: *renewable* and *sustainable*. Renewable refers to an energy source which is not depleted within reasonable time, whereas sustainable refers to the consequence of using the renewable energy source. A sustainable energy production should not decrease the possibility for future generations to enjoy the same standard of living as we (Europeans) enjoy today. A renewable energy source can therefore well be non-sustainable. Using wood for heating in urban areas can indeed be considered as renewable but may lead to air pollution and a range of diseases which can increase in magnitude with generations. Another issue is the concept of *reasonable time*. Clearly, an energy source lasting a few tens or a hundred years is not renewable. However, an energy resource which has the potential to deliver for a thousand years can be considered renewable, since the global development on such timescale is completely undetermined compared to the present.

With these definitions in place, we are in position of describing which energy resources can be labeled as renewable. Clearly, nuclear fusion belongs to this class, but also nuclear energy based on thorium or ^{238}U, as discussed in the previous chapter. The non-nuclear energy alternatives will be considered in this chapter. By order-of-magnitude, the largest energy resource is solar energy.[1] Solar energy is transformed into wind energy and hydropower. Solar energy also drives the biosphere and all bio-based energy forms. In addition, the wind causes a third generation energy resource at sea: wave energy.

[1] Solar energy is nothing but an end product of nuclear fusion taking place on the sun.

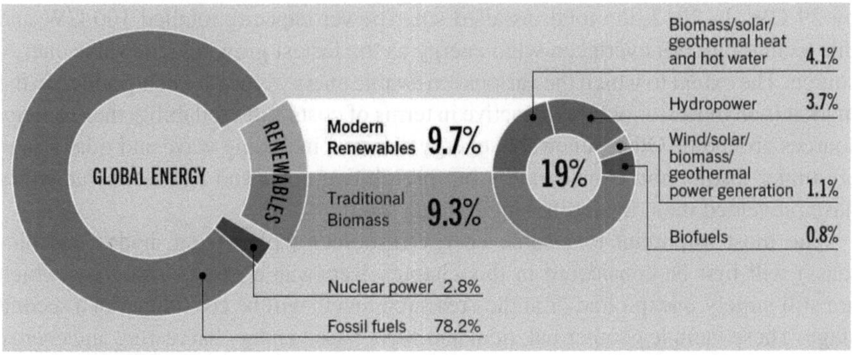

Fig. 4.2 Renewable energy share of global final energy consumption, 2011, © REN21 Renewables 2013 Global Status Report

Some comments on the oldest energy source of humans, biomass, are needed before entering this chapter. Up to this chapter, and in available statistics in general, the use of biomass for heating has not been included. The use of traditional biomass is however estimated to about 10 % of the world's primary energy consumption, and will be taken into account in this chapter. Biomass has been the dominant energy source up to the 18th century. Before the fossil era, wind energy was also extensively used for transport and for milling grain by wind mills. Solar energy used for heating has also been harvested by humans since ancient times. As such, turning to renewable energy sources is, in a sense, a return to our original energy. However, present technology allows for a more efficient transformation of the energy into work by several orders of magnitude.

As of 2011, the share of renewable energy in the global final energy consumption reached 19 % (REN21 2013), up from 16.7 % in 2010 (REN21 2012), with traditional biomass accounting for almost half this number, see Fig. 4.2. Hydropower dominates the other renewable resource in terms of importance. Concerns about climate change, increasing fossil fuel costs, decreasing fossil fuel reserves, decreasing renewable energy costs and potential to improve energy security are some of the elements that explain why more and more countries invest in renewable energy.

Yet, not all renewable energy types can serve the same purpose nor have the same degree of maturity or cost. Renewable energy can basically be classified in three categories: renewables for transport, renewables for electricity and renewables for heat. Biofuels such as biodiesel and bioethanol might compete, or add to, fossil fuels in the transportation sector. Hydro, geothermal, wind, modern biomass, concentrated solar power and solar photovoltaic might become substitutes to coal and natural gas for electricity generation. Traditional and modern biomass, geothermal and solar thermal are energy soures that can be used for heat and/or electricity generation.

To date, renewable energies are particularly striving in the electricity generation sector which increased by 213 GW globally in 2012. Almost half of this came from renewable energy: wind increased by 45 GW, hydro by 30 GW and solar PV by 29 GW. In 2012, the total installed solar power capacity totalled 100 GW and this technology has overtaken wind energy as the fastest growing renewable energy source. The extent to which the various renewable energy types are being added to the market is an indicator of how attractive in terms of costs and availability these energy sources are today. Other renewable energy sources, including wave and tidal power are under testing and consideration but prohibitive costs and technical challenges have prevented them from diffusing to large scale.

The most important renewable energy resources (solar, wind, hydro and biomass) will first be considered in this chapter. Renewable energy resources which are still largely unexploited or at the "research level" will be considered in a second stage. These include geothermal, tidal and wave based energy harvesting and energy exploitation based on salt gradients within sea water.

4.2 Solar Power

Solar power has a long history as energy source for humans. For example, solar power was used for heating of water in the Roman empire. A steam engine based on solar power was constructed by Auguste Mouchout in 1861, but was found to be far too expensive to have a commercial value. Modern solar power history can be said to start out in the 1970s as a result of the oil embargo. Now, 40 years later solar power remains a small part of the world energy mix. However, since 2011, solar power is the fastest growing source within the renewable sector.

Looking back at the power of the solar radiation it is clear that the present annual global energy consumption on Earth amounts to less than an hour of the total solar radiation which hits our planet. Thus, if we were able to transform just a tiny fraction of the total solar radiation into useful energy forms, the energy problem would be solved—once and for all. However, the problem with this seemingly simple solution is at least twofolds:

1. It requires large areas turned into powerplants and
2. it is, at present, much more expensive than fossil and nuclear alternatives.

We will start by discussing the three basic processes in use and consider area and cost requirements afterwards.

4.2.1 Basic Processes

Three processes have been implemented in practice to transform the solar radiations into energy: Solar photovoltaics (PV), passive solar power (PSP) and concentrated solar power (CSP). Solar PV and CSP are generally intermittent, non-dispatchable electricity generation sources. Combining solar PV with a battery or PSP/CSP

Fig. 4.3 Three technologies for energy conversion from solar radiation. *Left* A passive solarwater heating panel. *Middle* Concentrated solar power system. *Right* Photovoltaic cells on a residential house

with thermal storage permits us to produce potential energy. Both technologies are considered clean and renewable.

The idea behind PSP and CSP is very simple: The energy of the solar radiation is transformed into heat which increase the temperature of a fluid, usually water. The heated hot fluid may then be applied for direct heating of buildings or, in the case of CSP, as a working fluid in an electricity generating turbine.

- PSP is utilized by a black absorbing material which encompasses channels for water (see Fig. 4.3). The black surface absorbs almost all the incoming solar radiation which amounts from 100 to 500 W/m^2 at daytime depending on lattitude. Cloud coverage may change the numbers to the worse significantly. The total power absorbed, P_{in} depends on the area facing the sun and the efficiency of the conversion process. At constant conditions a part of this power is emitted due to the temperature difference between the absorbing plate and the environment. Then, with incoming solar flux I_{sun} and plate area A we have

$$\begin{aligned} P_{in} &= c_1 A I_{sun} \\ P_{out} &= c_2 \Delta T \end{aligned} \tag{4.1}$$

where c_1, c_2 are material constants and ΔT is the temperature between the plate and the environment. The efficiency of the process can then be defined as a dimensionless number between 0 and 1 as,

$$\epsilon_{PSP} = \frac{P_{in} - P_{out}}{A I_{sun}} = c_1 - \frac{c_2}{A I_{sun}} \Delta T \tag{4.2}$$

Note from this equation that when the incoming flux is large enough compared to c_2, the efficiency does not depend much on the last term. In practice it does and therefore the efficiency decrease when the temperature difference increase. Thus, solar collecting plates are most efficient when the temperature difference is small which implies that the temperature increase of the fluid is limited, in practice to something between 0 and 50 degrees compared to the outside temperature. Thus, far from equator only lukewarm water can be produced by this technology which results in an energy quality of the water which has limited applications.

- CSP is a technique to increase the conversion efficiency by increasing the incomming flux to the medium to be heated, i.e. to decrease the fraction $\frac{c_2}{I_{sun}}$ in Eq. (4.2). By constructing mirrors which focus all the incoming radiation towards a small concentrated spot containing the fluid, the conversion efficiency per mirror area can be made almost proportional to the constant c_1 in Eq. (4.2). In practice it is expensive to construct such perfect mirrors so a fraction of the radiation energy, often of the order of 50 % is lost in the mirror system. The temperature of the working fluid is nevertheless easily brought to many hundred degrees. This much higher fluid temperature can be combined with a turbine or engine system to produce electricity.

By assuming 100 % efficiency of the fluid to absorb the radiation, cf. Eq. (4.2), the efficiency of the electricity production process then depends in general on two factors: The efficiency of the mirrors and the efficiency of the engine which converts heat to electricity,

$$\epsilon_{CSP} = \epsilon_{mirrors} \cdot \epsilon_{engine} \tag{4.3}$$

Both factors depend on temperature of the environment and the working fluid respectively. A rough rule of thumb overall states an efficiency of 50 % for each term, which then results in a 25 % total efficiency. Both estimates may be higher, especially $\epsilon_{mirrors}$, leading to corresponding higher efficiencies.

- PV The principle behind photovoltaics is very different from the principle behind PSP and CSP and we have to go back to Chap. 1 to understand the physics. The solar radiation is nothing but a large bunch of photons with an intensity distribution given by the Planck curve. When a material is hit by the radiation, a given temperature dependent fraction of the photons have energy enough to excite electrons of the atoms which constitute the material. Certain materials (semiconductors) have the property that when their atoms are excited, the excitation energy (i.e the photon energy) can be turned directly into electricity. Now, since energy levels in materials are quantized, only photons with the correct energy corresponding to the energy difference between the initial (E_i) and final state (E_f) of the electron

$$hf = E_f - E_i \tag{4.4}$$

can contribute. The efficiency of PV cells is then essentially dependent only on the PV material, which governs the probability for electrons to be excited into a current supporting quantum state. Secondly, the PV efficiency also depends weakly on temperature since the number of photons available for excitation depends on the temperature, T and cloud coverage fraction, C.

$$\epsilon_{PV} = \epsilon_{material} W(T, C) \tag{4.5}$$

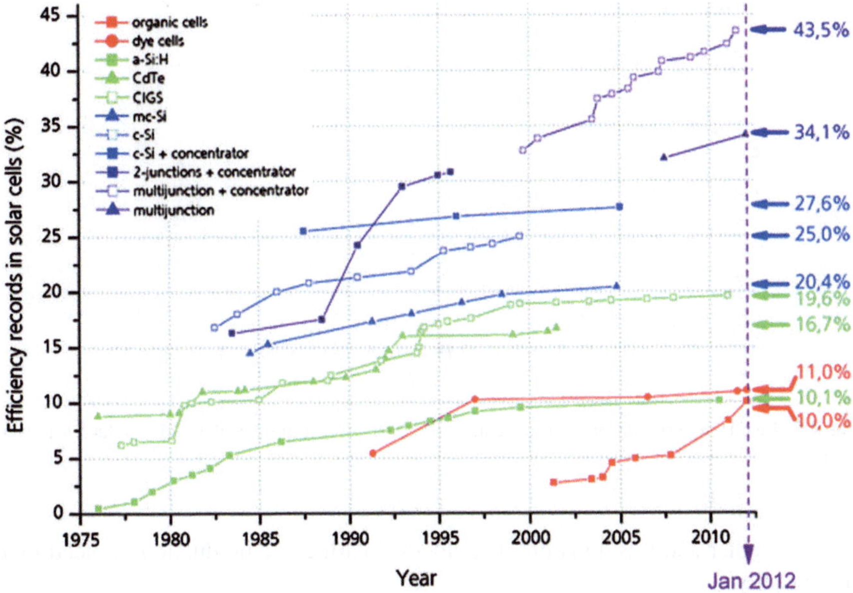

Fig. 4.4 Efficiency history of various PV technologies

where the temperature and cloud coverage fractions are given by the function $W(T, C)$. Ideally, the material has a range of excited energy levels $\{E_f\}$ which fits into Eq. (4.4) but in reality only a narrow band of final state energies exists. For the leading Silicon (Si) based materials of today the efficiency is around 20 % on a standard sunny day. More advanced, but also much more expensive, materials can be designed to reach an efficiency towards 40–50 % (Fig. 4.4).

Much less efficient, but cheap, organic materials offer efficiencies of a few percent. In Fig. 4.6 the efficiency of various materials are shown. We observe that the efficiency is in general increasing with time—since better insight to the physical processes in combination with enhanced production technologies (learning curve) improve. However, production costs for the most efficient materials are still such that they have no commercial value. Only one thing can be said for sure, regarding the future of the photovoltaics compared to CSP and other energy technologies: If a solar panel "Gore Tex" material is found, the finder will become incredibly rich!

The electricity production profile over four days of a large solar PV power plant (1 MW) is given in Fig. 4.5. The diurnal/nocturnal cycle can clearly be seen on the figure: solar PV only produces during day time. Then, the first day is a cloudless day and electricity production increases until it reaches a peak, before it decreases again. The production during the following three days is strongly affected

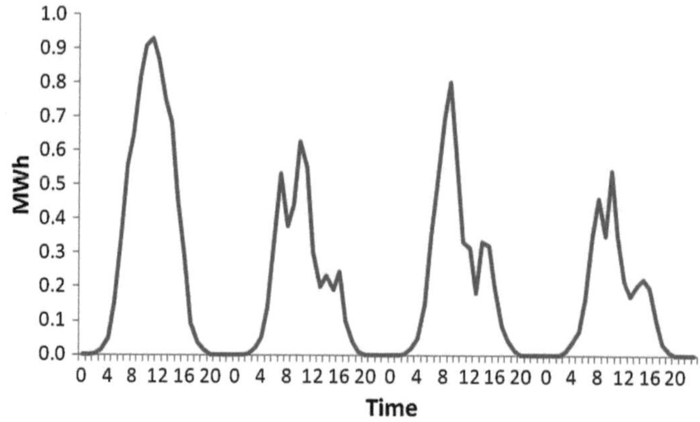

Fig. 4.5 Electricity production profile on an hourly basis over four days of a 1 MW solar PV power plant

by the weather and it is apparent that clouds do affect the production of electricity significantly.

The most common type of PV is the silicon-wafer-based PV. With high efficiency rates (typically between 14 and 22 %), these solar panels are traditionally being used on space-constrained rooftops. This technology was one of the first to be developed by the solar industry and is often referred to as a first generation technology. Silicon-wafer-based PV represented about 80 % of the market in 2010 (EPIA/Greenpeace 2011).

The main material necessary for the production of silicon-wafer-based PV is solar grade silicon or polysilicon. Silicon is found in sand and quartz and it is processed to reach a 99.99 % degree of purity. The basic material is abundant and prices are principally determined by the ability of the industry purifying the silicon to meet the demand. The solar industry is mostly competing with the electronic industry for this resource, the former being the main user of that material since 2008. The second type of photovoltaic technology and often referred to as second generation technology, is the thin-film photovoltaics. As the silicon-wafer-based technology, the thin-films are usually installed on rooftops. Thin-films are 'cheap' compared to the other types of photovoltaic as the cost is roughly half the cost of the silicon-wafer-based technology. On the downside, the efficiency rates are much lower (4–12 %). Thin-films will often be choosen if the space available is not constraining the size of the installation.

Several types of thin-film PV technologies exist, which rely either on semiconductors like Silicon (Si), Cadmium (Cd) and Tellurium (Te) or on metals like Copper (Cu), Indium (In) and Gallium (Ga). The technology relying on Cd and Te is dependent respectively on the zinc and copper mining industries as these two materials are byproducts of the Zinc and Copper extraction (EPIA/Greenpeace 2011). The long-term availability of these materials might become an issue. No sign of shortage is foreseen for the other primary materials.

Silicon-wafer-based PV and thin-films PV can either be installed as residential or commercial systems or as utility-scale power plants. In residential and commercial systems, these technologies are either mounted on top of the roof or on the facades. The energy generated by these technologies will primarily serve to cover the needs of the owners. If there is excess generation and if the installation is connected to the electric grid, the owner will be able to sell the excess power. In utility-scale applications (bigger than 1 MWp), the space requirement is much bigger. Therefore, such installations will often be centralized and ground-mounted. The quasi-totality of the power generated will be sent into the grid.

4.2.2 Area Requirements

Independent of type of solar energy conversion technology it is clear that it is limited by the incoming radiation amounting to more than 400 W/m^2 on average at the equator and decreasing to about half that amount in central Europe. In Germany, a PV based power plant, rated at 15 % efficiency can then be expected to deliver 20–30 W/m^2. Realistically, lets half this number,[2] to take into account cloudy days, maintainance, etc. With 10 W/m^2 a 1 GW power plant requires an area of 100 km^2. This is a large area, 10 km in each direction of flat ground transformed into a powerplant system which excludes, farming, houses or infrastructure—only solar collectors all over. We then understand that, in habited areas, solar energy is difficult to implement "at full power". Small scale power production by covering buildings with PV plates for example, is however still a posibility. On the other side, large areas are available in sunny unpopulated regions all over the world. The six circles of the map in Fig. 4.6 would produce about 18 TW of electric energy, more than the present annual energy consumption on Earth.

4.2.3 Present Use, Resource Considerations and Forecast

Concentrated solar power is the closest form of solar energy compared to the conventional energy generation sources as we know them today. The efficiency levels reached commercialy, ranges between 25–35 %. Of the three technologies presented in this section on solar power, CSP is typically the cheapest. Four types of CSP are currently available in the market: parabolic trough, Frensel reflectors, solar tower and parabolic dishes (IEA 2010). To be competitive, CSP capacity needs to be placed in areas which receive over 2,000 kWh per square meter per year of direct solar radiations.

Over 1 GW of CSP had been installed by the end of 2010, with 740 MW built between 2007–2010 (REN21 2011). Nearly all the capacity has been built in Spain

[2] The Solarpark in Mühlhausen in Bavararia is expected to produce at 5 W/m^2.

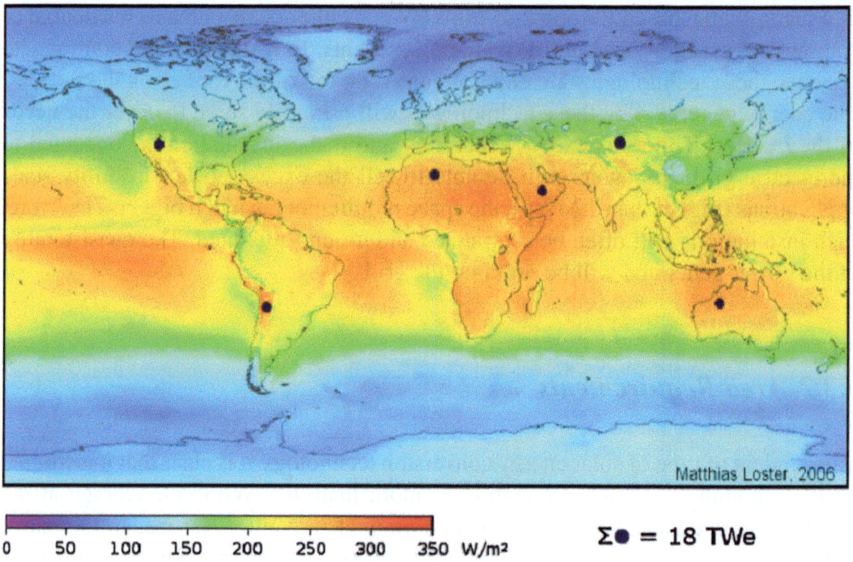

Fig. 4.6 The *dark blue circles* shows the land area needed to produce about 18 TW electricity (with 8 % efficiency). *Source* Wikimedia commons

and in the United States, even though many countries would be highly suitable for that type of energy. Countries in Northern Africa and in the Middle East are now considering concentrated solar power and the first projects should be built soon. The parabolic trough technology dominates the market.

Presently, two groups of solar PV compose the market: silicon-wafer-based PV and thin-film PV. Silicon-wafer-based PV systems are more efficient and costlier than thin-film PV. Therefore, silicon-wafer-based PV will be prefered in situations where the space available is constrained. It is only since 2002 that solar PV has massively been installed globally and at an increasing rate, see Fig. 4.7.

At the end of 2012, over 102 GW of solar PV had been installed in over 100 countries, with most of it being located in Germany, Spain, Japan and Italy. Of these 102, 31 GW have been installed in 2012 only.

Solar photovoltaic and concentrated solar power relied heavily on subsidized conditions to start their adventure. Learning effects, increased efficiency over time and a decrease in manufacturing costs taken together led to an impressive reduction in PV cells' cost. However, solar photovoltaic has not yet reached grid parity and thus still remains dependent on generous policy instruments to drive down its costs and support the expansion of this energy source. Grid parity occurs when the electricity generation cost of a new technology equals the cost of conventional electricity sources and such situation is predicted to occur soon in specific locations as PV modules will experience further cost reductions and as electricity prices are generally following an upward trend.

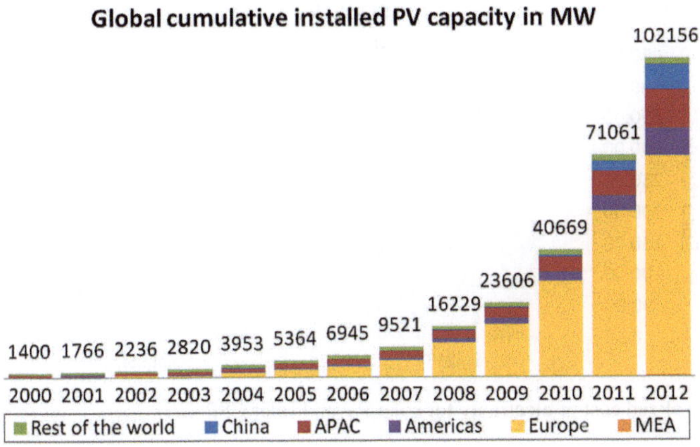

Fig. 4.7 Global cumulative installed photovoltaic capacity in MW, 2000–2012 (REN21 2013)

4.2.4 Cost

4.2.4.1 Resource Cost for Solar Power

Solar power is not subject to any fuel cost, since no price can reasonably be attached to solar irradiations. It implies that when a solar project is paid back for, there is a need only to cover the O&M costs with the rest of the revenues being profit. Also, the absence of fuel costs impact the risk levels of future costs as it suppresses an important category of risk, which conventional energy generating sources have to take into account.

4.2.4.2 Basic Cost of Solar Power

Solar PV and solar CSP are capital intensive technologies, since the capital costs account for between 80 and 90 % of the total cost of a solar installation over its lifetime. A solar PV system is basically composed of PV panels, which main components are PV modules (device containing 36 PV cells), and of the balance of system (BOS), which encompasses all the elements of a PV system other than the panels. The capital cost of such system can amount to between 2,000 and 5,000 Euro/kW. For solar CSP projects, the solar field in itself often accounts for a third of the capital cost and the thermal system for most of the remaining capital costs, which lie between 3,000 and 3,500 Euro/kW of installed capacity. Operation and Maintenance costs account for the remaining 10–20 % of the levelized costs. For solar PV, O&M costs include scheduled maintenance such as the cleaning of the panels to ensure maximum efficiency, non-scheduled maintenance and the replacement of inverters, which have a tendency to fail (Jacobi and Starkweather 2004). O&M costs associated

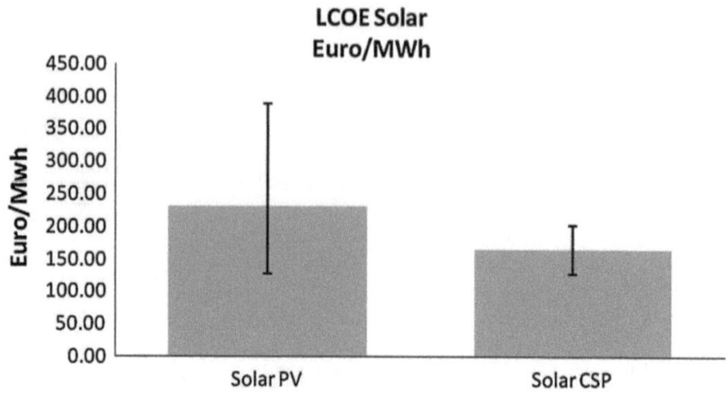

Fig. 4.8 Levelized cost of electricity for solar PV and solar CSP

Table 4.1 Levelized cost of electricity for onshore and offshore wind power

	Solar PV	Solar CSP
LCOE (Euro/MWh)	128–389	127–202
Overnight cost (%)	80–98	85–90
O&M costs (%)	2–20	10–15
Fuel costs (%)	0	0

to solar CSP are very similar to those of other steam power plants. A particularity of a CSP plant is the need for periodical cleaning of the field mirrors.

4.2.4.3 Electricity Generation Cost of Solar Power

The electricity generation cost of PV depends heavily on the solar irradiation levels and solar PV devices will therefore be comparatively cheaper in the Middle East compared to northern Norway or central Europe. Estimates from the authors under the assumptions used to evaluate the levelized cost of all energy technologies, places the levelized cost of solar PV between 13 Eurocents/kWh and 34 Eurocents/kWh and the levelized cost of solar CSP between 13 Eurocents/kWh and 17 Eurocents/kWh (see Fig. 4.8, Table 4.1).

Besides the solar irradiations levels, the amount of storage can significantly impact the economics of a CSP plant, since storage availability results in an increase in the annual capacity factor and thus improve the economics of a plant.

Is Solar Energy Really that Expensive?
Comparing technologies purely based on their LCOE shows that solar PV is five times more expensive than coal on average. Yet, this comparison is somewhat unfair to both technologies.

First, the electricity produced by solar PV is often consumed by the household which has installed these panels on its roof. Using this electricity allows the household to avoid paying for electricity transportation costs (i.e.: the grid) which can often be equivalent or exceed the cost of electricity. Hence, assuming that the household is located in a sunny area, the cost per kWh from PV can be as low as 13 Eurocents/kWh. Buying the electricity on the grid, in the other hand would cost the household about 4–6 Eurocents/kWh. Double this number for the cost of transportation and add taxes and it could be that it is cheaper for the household to get power from its solar panels than from the grid.

Second, solar PV do not generate power all the time. For example, during winter months in Nordic countries, the lack of sun is likely to prevent any production of electricity at all, hence a 100 % solar PV based electricity system is unrealistic, unless a significant amount of storage is built (which, in itself, is likely to make the LCOE of solar PV + storage much higher than the LCOE of just solar PV). Consequently, if the electricity from solar PV may appear cheap to some, it is principally because they compare two technologies (solar PV and others) in an unfair manner, often forgetting the cost of intermittency of solar PV. So, is solar competitive? Yes, but only under strict assumptions.

Yet, the cost of solar PV is decreasing at a rapid pace as we learn how to harvest solar radiations more efficiently. It could be that in sunny areas with relatively high electricity prices (Southern Italy for instance), building solar panels on one's house will make financial sense even without subsidies in a near future.

4.3 Wind Power

The process of transforming wind energy into mechanical energy has, as for solar energy, been known for a long time. For example, the use of wind energy for sailing goes back several thousand years in human history. In more recent times, between 150 and 200 years ago, Europe had about ten thousand operating windmills which were used to grind grain and to raise water for irrigation for example. The peak power of these windmills, let's assume 1 kW, seems tiny compared to the nowadays peak power of 4–5 MW of capacity of modern wind turbines.

Wind energy is second generation solar energy, because the wind is caused mainly by energy absorption of incoming solar radiation in the atmosphere. The temperature gradient between the equator and the poles are then causing two major

Fig. 4.9 Photo of an old windmill in Aarhus, Denmark. (*Photo* Stan Shebs, August 2003. *Source* Wikimedia Commons)

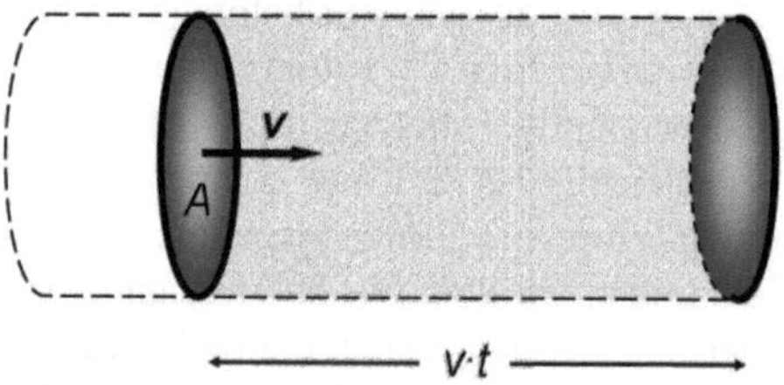

Fig. 4.10 Illustration of the mass–volume relationship: A parcel of wind with cross sectional area A and length vt to pass through a wind turbine within time t

northern and southern large scale wind circulation belts (the sub-tropical jets or the Westerlies). The Earth's rotation combined with the distribution of land and sea areas are additional important mechanisms for the creation of wind patterns. For example, the origin of the almost constant drift of low pressures systems in the north-eastern direction across the Atlantic hitting western Norway with never-ending rain periods, is due to the combined effect of the Rocky Mountains and land-sea temperature contrasts. The Rockies appear as an obstacle to the sub-tropical jet and this creates waves and disturbances in the flow downstream of the mountain range. As the jet leaves the North American continent and enters the North Atlantic, it experiences thermal contrasts that amplify the disturbances. These grow into eastward traveling cyclones that constitute the North Atlantic storm track, often ending up on the Norwegian coast (Figs. 4.9 and 4.10).

4.3.1 Basic Processes

Energy conversion in wind turbines is made possible by the kinetic energy of the wind molecules colliding with the rotor blades. This process involves a reduction in the speed of the wind to the benefit of increasing and maintaining the rotational energy of the rotor. Consider a parcel of wind with a given mass m and velocity v (see Fig. 4.10). The kinetic energy of the air parcel is $\frac{1}{2}m \cdot v^2$ and since the mass is the product of the volume of the parcel volume V and density ρ we can write:

$$E_{wind} = \frac{1}{2} \cdot \rho \cdot V \cdot v^2 \tag{4.6}$$

The mass of the wind parcel is replaced by $m = \rho \cdot V$ in Eq. (4.6). The volume of the wind parcel, V, is equal to its cross sectional area times the arbitrary length $l = v \cdot t$ considered, thus:

$$E_{wind} = \frac{1}{2} \cdot \rho \cdot A \cdot vt \cdot v^2 \tag{4.7}$$

The total wind power, $P_{wind} = E_{wind}/t$ passing the area swept by the wind turbine can then be expressed as:

$$P_{wind} = \frac{1}{2} \cdot \rho \cdot A \cdot v^3 \tag{4.8}$$

Note that the most crucial factor affecting the total wind power is the wind speed, because the wind speed contributes by a cubic dependence. Doubling the wind speed consequently gives eight times more power. In addition, the power depends linearly on the density of the wind, which can be assumed constant about 1 kg/m^3 anywhere on the surface of the Earth. Finally, the power increases linearly with the cross sectional surface area of the rotor blades. The latter element implies that the power hitting a turbine with rotor radius r_1 is twice the power hitting two turbines with rotor radius $r_{1/2}$ since $A = \pi \cdot r^2$.

We all know that wind speeds vary on average from place to place. During a fine summer day, you may find that cycling in one direction is almost without resistance while it is much more difficult to keep up the speed against the wind. This indicates that the wind speed is, in order of magnitude, equal to the cycling speed, for example 5–7 m/s. During a stormy day, the wind speed may increase ten folds and cycling in any direction would not be recommendable. Thus, the average and the variation of the wind speed are crucial for the power which can be extracted. 5–7 m/s is in fact the expected average wind speed of many relatively windy places in the world (e.g. most costal areas of western Europe). Average wind speeds may be lower inland, typically around 3–5 m/s in Europe. Some costal regions, in particular offshore regions, may have up to 8–10 m/s average wind speeds.

In addition to different average wind speeds, we know that at a specific spot, wind speeds vary strongly with time and the daily, weekly and seasonal varia-

Fig. 4.11 Illustration of the daily wind variation measured each 5 min at a weather station in England for October 21st, 2013

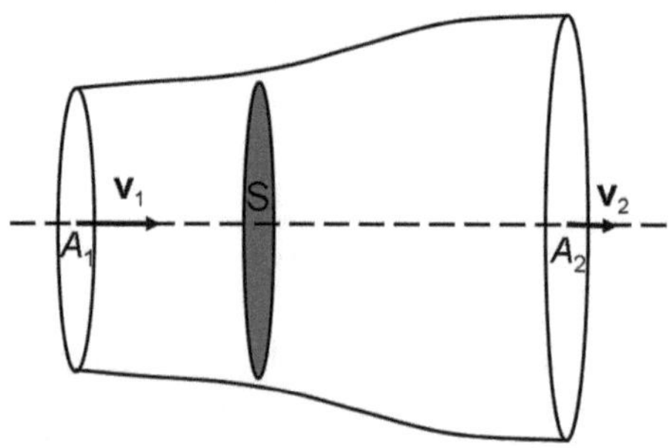

Fig. 4.12 Volume expansion of a wind parcel before and after it passes a wind turbine at S

tion may be large. Figure 4.11 illustrates how wind speed varies within a 24 h period.

There are two fundamental limits to the efficiency of a wind turbine. The first is called *Betz law* (Betz 1966) and is related to the fact that the entire energy of the wind parcel cannot be absorbed by the turbine. A total absorbtion of the energy of the wind parcel would cause the wind to a "standstill" behind the rotor, prohibiting more wind to pass. Let us therefore assume that the wind passing through the rotor area can be schematically illustrated as in Fig. 4.12.

The wind flows through an area A_1 with velocity v_1 before the rotor area (we set the rotor area to be equal to A_1). After its passage through the rotor area, the wind vanishes through the area A_2 with velocity v_2. Note that $A_2 > A_1$ as long as the density of the air is roughly constant. The wind velocity at the rotor can be approximated by $(v_1 + v_2)/2$. The power delivered to the wind turbine is:

$$P_{wind} = \frac{E_1 - E_2}{t} = \frac{1}{2} \cdot \frac{m}{t} \cdot (v_1^2 - v_2^2)$$
$$= \frac{1}{2} \cdot \frac{\rho \cdot V}{t} (v_1^2 - v_2^2) = \frac{1}{2} \cdot \rho \cdot A \cdot v \cdot (v_1^2 - v_2^2)$$
$$= \frac{1}{4} \cdot (v_1 + v_2) \cdot (v_1^2 - v_2^2) \tag{4.9}$$

The efficiency of the wind turbine is the fraction of absorbed power in the rotor blades and the incoming power $P_1 = E_1/t = 1/2 \cdot \rho \cdot A \cdot v_1^3$:

$$\epsilon = \frac{P_{wind}}{P_1} = \frac{1}{2} \cdot (1 + x - x^2 - x^3) \tag{4.10}$$

where the last equation appears when we introduce $x = v_2/v_1$. Then, the maximum possible efficiency of a wind turbine is found by differentiation. We obtain:

$$\epsilon = 59\ \% \text{ for } x = \frac{1}{3} \tag{4.11}$$

Hence the maximum efficency of a single wind turbine, independent of wind velocity is about 60 %. In practice, a wind turbine also needs some maintenance. It is also known that the turbine system has a maximum power production, which implies that the efficency drops towards zero for wind speeds above 20 m/s, speeds at which a "theoretical" wind turbine becomes superefficient. Finally, the efficiency of the conversion process from mechanical energy of the rotating blades into electricity has to be considered. This efficiency factor is normally quite high, around 90 %. In total, this gives a reasonable estimate for the power conversion efficiency of around $\epsilon_{total} \sim 30 - -40\ \%$.

The second fundamental limit is related to the fact that the wind diverges into a larger flow area behind the wind turbine (see Fig. 4.12). That implies that wind turbines that are placed too close to each other will give less power production compared to two independent wind turbines. To illustrate this fact, imagine two wind turbines located behind each other in the direction of the wind. The second wind turbine experiences the wind which has been slowed down by the first one. Since the slowed down wind also expands to a larger area behind the first row of wind turbines, it is clear that a certain distance between the wind turbines in the direction perpendicular to a given wind direction is necessary in order to allow for areal expansion of the wind pattern behind all "first row" wind turbines to disappear. Within the industry, it is found that a distance equal to five times ($d = 5$) the rotor blade diameters is necessary between each wind turbine to reach the full efficiency for all wind turbines. The power per area can now be obtained by replacing A by the power per wind turbine $\epsilon_{total} \cdot \pi \cdot \left(\frac{d}{2}\right)^2$ in Eq. (4.8) and dividing by the land area per wind turbine $(5 \cdot d)^2$ gives:

$$p_{wind} = \frac{\frac{1}{2} \cdot \rho \cdot \epsilon_{total} \cdot \pi \cdot \left(\frac{d}{2}\right)^2 \cdot v^3}{(5 \cdot d)^2} = \frac{\pi \cdot \epsilon_{total} \cdot \rho v^3}{200} \tag{4.12}$$

Fig. 4.13 *Left* Lillgrund Wind Farm's wind turbines in the Sound near Copenhagen and Malmö (Mariusz Paździora, Wikimedia Commons) *Right* Onshore wind farm in Ardrossan, North Ayrshire, Scotland (Vincent van Zeijst, Wikimedia Commons)

Table 4.2 Power per area for a wind farm with average wind speed $v = 6$ m/s

Efficiency	Power per area (W/m^2)	Wind speed (m/s)
0.3	1.25	6
0.4	1.65	6
0.4	3.92	8

Note here that the power per area is independent of rotor blade radius. Smaller wind turbines allow for more wind turbines per area, which compensate for the smaller power production per wind turbine. However, larger wind turbines are driven by wind at an average higher altitude than smaller ones, and the average wind speed increases in general with altitude. In Fig. 4.13, images of two different wind farms are shown. Note the distance between each wind turbine!

4.3.2 Potential and Area Requirements

By using Eq. (4.12) with a typical wind speed of $v = 6$ m/s gives a power per area p_{wind} of 1–2 W/m^2, depending on precise efficiency values (see Table 4.2).

Assuming the lower limit for land area wind farms, a 1 GW system requires an area of about 1,000 km^2, which is ten times larger than the area needed for a solar farm of similar capacity. In addition to price comparison (as we will see soon), wind turbines have another advantage over solar panels: wind turbines do not exclude other land area uses, such as farming or even additional covering of the ground with PV solar cells.

There are also regions where the wind speed on average is larger than $v = 6$ m/s. Many regions along shores or even at sea, for example in the western parts of Norway and in the North Sea, have average wind speeds around $v = 10$ m/s, which translates to a power per area of about 7 W/m^2. Thus, the power production in windy areas from wind energy is close to the expected power production per area of solar radiation in sunny areas.

According to Fig. 1.6, about 20 % of the incoming solar radiation is absorbed by our atmosphere. Using this number, it is possible to estimate the potential of wind energy by remembering that almost 90 % of the mass of the atmosphere amounts to a 10 km high slab of air above the Earth's surface and that the part of the wind which can be tapped into is located near the surface, i.e. up to 100 m above ground levels. Assuming further that the wind energy is uniformly distributed up to 10 km, we can estimate an order of magnitude for the total wind power potential, which is accessible to us.[3] A fair share (about 50 %) of the absorbed energy in the atmosphere is converted into kinetic energy of the atmospheric molecules. This gives a total available power of:

$$P_{total}^{wind} = 50\,\% \cdot 0.9 \cdot \frac{0.1\text{ km}}{10\text{ km}} \cdot 0.2 \cdot 174000\text{ TW} = 156\text{ TW} \tag{4.13}$$

This number is remarkably close to the correct wind energy potential of 130 TW calculated by Gustavson (1979).

This power is distributed unevenly at all times around the surface of our planet, and clearly only a small fraction of it can be utilized. Many areas, including urban areas, arctic regions, high altitude regions and similar due to technical and/or political constraints have to be excluded from the area available for extracting power from the wind. An upper limit of the economic potential of wind energy is realistically only a few percent of the Earth land area. Using a modest 2 % land area results in a potential of about 3 TW. Opening up for offshore wind power, the global wind potential would reach around 5 TW.

Many studies have aimed at estimating the 'true' potential of wind power. For example, Archer and Jacobson (2005) analyzed over 7,500 surface stations and another 500 ballon-launched stations in an attempt to evaluate the mean annual wind speed at a height of 80 m. They concluded that the world wind resources may be sufficient to technically (albeit not necessarily economically) meet with current final energy demand. In a more conservative analysis, Hoogwijk et al. (2004) calculated the technical potential of global onshore wind energy to lie between 2 and 6 TW, which is more in line with our estimate. These studies show that wind energy, as solar energy, also have a significant resource potential. However, in the perspective of a future 30–60 TW total world power level, wind will play a limited role.

4.3.3 Present Use, Resource Considerations and Forecast

The global installed wind capacity has been multiplied nearly 46 times between 1996 and 2012 to reach 283 GW of installed capacity at the end of 2012. It is equivalent to about 200 to 500 coal power plants. Including intermittency in the picture, the existing

[3] Could you imagine a wind turbine of 1 km in height with almost the same length of rotor blades?

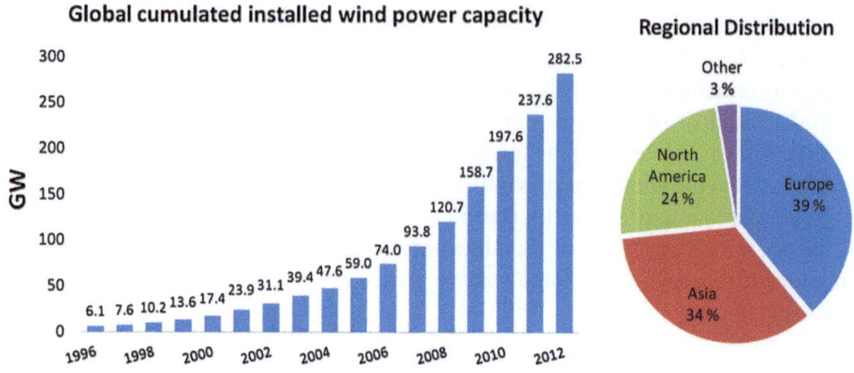

Fig. 4.14 Global cumulative installed wind capacity and wind capacity distribution in 2012 (GWEC 2013)

wind turbines generate as much electricity as 100 coal power plants.[4] China is leading the world in terms of installed capacity, in front of the United States and Germany.

Along with an increasing popularity around the world, wind turbines seen today differ significantly from the turbines available twenty years ago. Two types of changes are visible. First, wind turbines grew taller over time, which allowed them to reach stronger winds and reduce the land effect, i.e. the wind slowing down near the ground due to obstacles such as trees for example. Second, the wind turbine diameter has also increased overtime, meaning that the area swept by the blades has expanded and more energy from the wind can therefore be captured by a single wind turbine. Between 1990 and 2007, the average power of the wind turbines sold on the market jumped from a mere 200 kW to an impressive 2 MW. Even larger wind turbines with a rated capacity of 5–7 MW are being tested today and wind turbines from the range of 10–20 MW are being evaluated. If taller and larger wind turbines mean that more energy from the wind can be extracted from a single wind turbine, it also means that the structure needs to be strong enough to support the weight of the nacelle, rotor and blades. Thus bigger does not necessarily mean better from an economic perspective and bigger wind turbines have so far been restricted to offshore sites, where the incremental cost of installing larger wind turbines is offset by lower installation costs.

Most of the wind turbines installed to date are generating electricity, which is sold to the industrial and residential sectors via the grid. Some cases of stand-alone electricity production from wind turbines and some water pumping applications also exist.

Relying mostly on local wind conditions, wind power is far from having reached its full theoretical potential, limited only by the size of the industry, the technology available and especially, the economics of wind power. In some countries, including Denmark, wind resources are limited and most economically viable locations have

[4] 85 % plant availability is assumed for coal-fired power plants and 30 % capacity factor for wind power.

Fig. 4.15 *Left* Location of the planned Dogger Bank offshore wind farm (*Picture* NASA). *Right* mounting of the Statoil Hywind offshore wind turbine outside Karmøy, Norway in 2009 (*Picture* Jarle Vines, Wikimedia Commons)

been filled. Therefore, new installations often replace older and less efficient wind turbines. Further development, implying installing wind turbines on less favorable locations will only be achieved if subsidies are increased, if the cost of wind power is further reduced and/or if the electricity prices keep rising. In other countries, including the United States, India and China, space is not constraining the development of wind power and it is typical to see less efficient turbines, but cheaper and in greater numbers.

If many believe wind power will gain in importance in the future, its deployment is conditional to how close to grid parity wind power will get in the future and on the development of technologies that can compensate for the intermittency issue. Recently, especially in countries which have exhausted most of their favorable onshore sites, the interest in harvesting wind resources offshore is growing. An illustration of this growing interest is the licence awarded in 2010 to develop 9 GW of offshore wind power at the Dogger bank site, see Fig. 4.15. If the full 9 GW is developped, this project alone will exceed the total capacity installed at the end of 2012 (5.4 GW of offshore wind power were in operation worldwide at the end of that year). In addition, if such projects become real, the type of wind turbine installed may be floating instead of standing on the ocean floor. An example of such tower is the Hywind 2.3 MW wind turbine, which is under testing outside Karmøy in Western Norway.

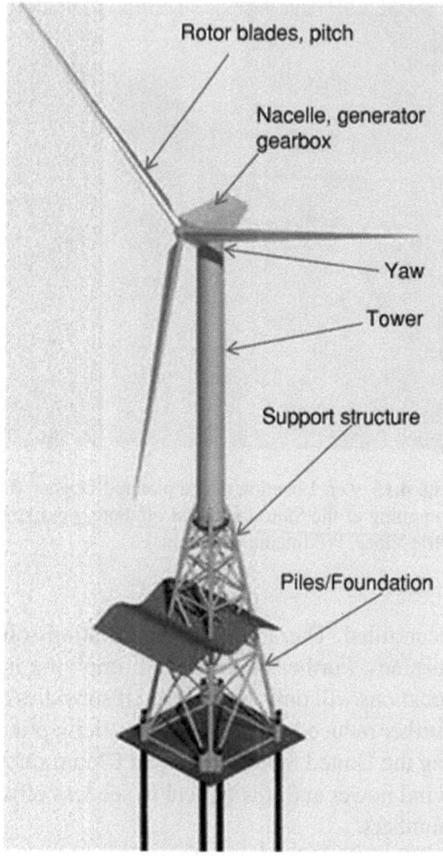

Fig. 4.16 Components of a wind turbine

4.3.4 Cost

4.3.4.1 Basic Cost of Wind Power

The central elements of a wind turbine are displayed in Fig. 4.16. A wind turbine basically contains a tower, a nacelle which includes the generator and a rotor system. The tower can reach over 150 m in height and needs to be constructed in a way that it can support the rotor system at the top. The nacelle can rotate to constantly face the wind and a pitch controller optimally orientate the blades with respect to wind direction in order to maximize the efficiency of the wind turbine for different wind conditions. The rotational energy is transformed into electricity in the generator which is coupled to an electric grid or a storage system.

Wind power is dependent on wind to produce electricity. As wind is a free comodity, no fuel costs need to be bared by the producer. Operation and maintenance costs in the case of wind power are relatively low compared to other electricity generating technologies. O&M costs are generally between 1 and 4 Eurocents/kWh and include various insurances, overhead costs, spare parts and labour costs. Generally, O&M costs increase with the age of the wind turbine but decrease (so far) with the year of installation of the wind turbine as newer models are better optimised for a specific location compared to older models.

In the absence of fuel costs and with low O&M costs, the capital costs are responsible for a big share of the total cost of a wind power project over its lifetime. The overnight costs per kW of installed onshore wind power capacity can vary greatly and currently lies between 800 Euro/kW and 2,500 Euro/kW (OECD/Nuclear Energy Agency 2010). The wind turbine in itself accounts for most of the total investment as it makes between 68–84 % of the total investment cost (Krohn et al. 2009). The cost of connecting a wind turbine to the grid accounts for another 2–10 %, while the foundations account for 1–9 %, the electric installation for 1–9% and the land rent for 1–5 %. The capital costs are predominantly influenced by the type of wind turbine being installed, its height, the distance from the grid and its location. For example, it can cost up to an additional 35 % to install a wind turbine in Canada compared to installing the same wind turbine in Denmark, because of the rougher conditions seen in Canada. Beware! It is not because wind power is more expensive to install in one region compared to another that wind power will necessarily be less attractive in that region, especially if subsidies and electricity prices vary between these two regions.

The figures are different for offshore wind power where the cost per kW of installed capacity is comparatively much higher as it currently lies between 2,000 Euro/kW and 5,000 Euro/kW. This comparatively higher price is because offshore wind power is not as mature compared to onshore wind power and because it is comparatively more challenging and expensive to install a wind turbine off shore (Edenhofer et al. 2011). Recurrent bottlenecks in the supply chain also contribute to raising the price of the technology, as well as volumes that are not significant enough to induce sensible cost reductions yet.

4.3.4.2 Electricity Generation Cost of Wind Power

Natural and man-made factors influence the cost per kWh generated by a wind turbine. The mean annual wind speed and the wind speed distribution at a site fall under the former category. Strong, but still within the boundaries of what a wind turbine can handle, and stable winds will ensure a maximal use of the turbine. Such conditions, added to an ability (i.e. man-made) to cheaply tap into that resource will ensure best returns. The type of wind turbine used is therefore of crucial importance here since not all models are adapted to every locations. For this reason, using different models on one site can result in very different electricity generations costs.

As mentioned previously, capital costs account for the big part of the levelized cost of electricity, see Table 4.3, and wind power is therefore a capital-intensive technology, because most of the cost of a wind farm over its lifetime occurs before

Table 4.3 Levelized cost of electricity for onshore and offshore wind power

	Onshore wind	Offshore wind
LCOE (Euro/MWh)	47–86	95–124
Overnight cost (%)	74–88	70–94
O&M costs (%)	12–26	6–30
Fuel costs (%)	0	0

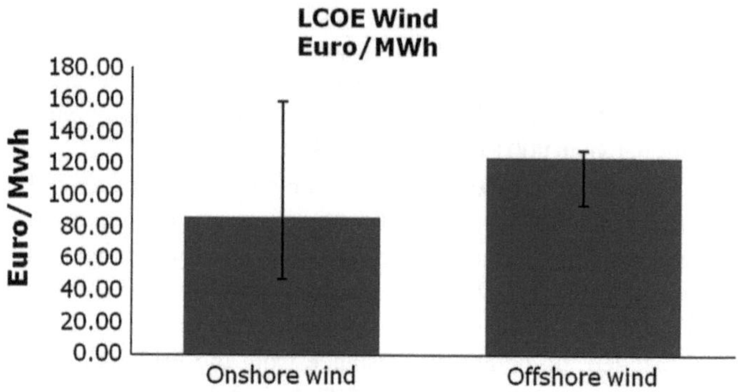

Fig. 4.17 Levelized cost of electricity for onshore and offshore wind power

any output is generated. Capital intensive means that the ability to secure access to cheap capital is a crucial factor for facilitating the installation of wind turbines over a territory. Such high shares of capital with respect to the total cost of a project over its lifetime implies that once the project is operational, there is little uncertainty in the future costs of a project compared to other energy sources. First, a wind project often consists of several wind turbines and the failure of one turbine does not prevent the others from functioning. Second, there is no uncertainty attached to the evolution of future fuel costs.

The main source of uncertainty attached to a wind project is the future wind conditions on a site which will affect the annual energy production from a wind power plant, and thus the level of income and of the variable costs of a project per unit of output.

In terms of costs, the minimum and maximum cost, as well as the median cost for both onshore and offshore wind technologies are illustrated in Fig. 4.17. The levelized costs of onshore wind energy amount to 4.7–8.6 Eurocents/kWh, whereas the costs for offshore wind energy are around 9.4–12.8 Eurocents/kWh. The sizeable range of costs for onshore wind power is the result of the impact of external factors from the technology in itself, with the location clearly being the main factor.

The role of the location and more precisely the impact of the wind regime on the cost of wind power is illustrated in Fig. 4.18. Favorable locations are those offering strong, but manageable and stable winds. A wind turbine installed in such locations will be able to generate more power and more often than a wind turbine placed in a location with weak and occasional winds. Consequently, the wind regime will impact the economics of a wind turbine.

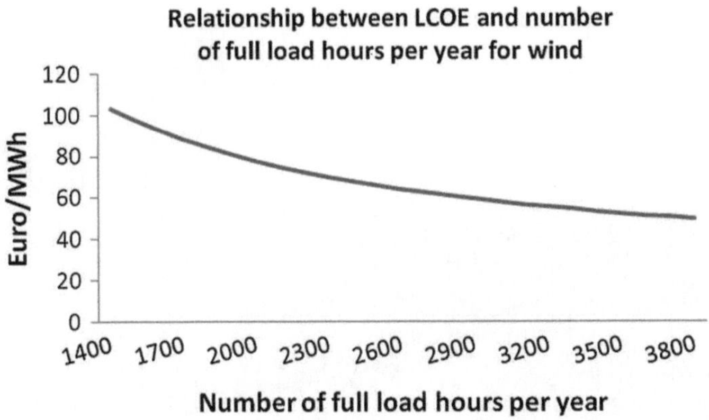

Fig. 4.18 Example of LCOE from wind per MWh as a function of the number of full load hours per year

A rated 3 MW wind turbine would be generating 26,280 MWh per year (3 MW · 24 h/day · 365 days/year) if it was operating all the time. At times, the power plant will not operate at full capacity, especially is winds are too weak/strong or if it is under maintenance. The element of interest in practice is the number of full load hours equivalent per year, which is the observed production divided by the maximum theoretical production. If a total of 5,256 MWh of electricity is generated in a year by our 3 MW wind turbine, the capacity factor is 20 % (5,256 MWh/26,280 MWh) and the full load hours per year amount to 1,752 (5,256 MWh/3 MW). Total costs have to be spread over the total quantity of electricity generated, and the higher this number, the lower the cost per unit of electricity generated by the wind turbine over its economic life. Figure 4.18 is based on the median case of onshore wind power and reveals the negative relationship between cost per unit of electricity generated and the number of full load hours per year. Higher or lower capital costs would shift the curve upward or downward respectively.

It can be concluded from Fig. 4.18 that the location is one of the key factor in determining the economics of wind power as the number of full load hours (and by extention, the capacity factor) will influence the cost of generating electricity from wind. As mentionned earlier in this chapter, wind speeds are generally weaker inland compared to coastal areas, mostly due to the presence of obstacles slowing down the wind as it makes its way inland. Some locations inland, especially in some particular valleys and in montainous regions, are also very favorable for installing wind turbines. In terms of capacity factor, onshore wind turbines can reach capacity factors of 20–30 % with a few examples approaching or exceeding 40 %.

Although offshore wind power is costlier than onshore wind power, offshore wind turbines generally benefit from stronger and more stable winds compared to onshore wind turbines. This fact is reflected in the capacity factor of many offshore wind turbines, which is often between 35 and 45 %. Stronger and more stable winds are

Fig. 4.19 *Left* Itaipu Dam on the border of Brazil and Paraguay, with a capacity of 14 GW. *Right* Untouched hydroelectricity: A Waterfall in Iceland

the main reasons why offshore wind power attracts a lot of interest, despite its high current cost.

Another key factor in determining the cost of wind power is the price of raw material. During the prices surge of raw materials between 2006 and 2008, the wind turbine cost followed the same trend and increased significantly.

4.4 Hydropower

Peforming work from falling water is a known process since at least 3,000 years BC (Fig. 4.19). Modern hydroelectric production via a turbine has been established in the late part of the 19th century. Hydropower has since then played an increasing role in modern industrialization and it is today a major global energy source and now contributes to about 16.5 % of the global electricity production (REN21 2013), very comparable to the share of nuclear power. The source of hydropower stems from the evaporation of sea water and cloud formation, which in turn, is a result of solar radiation. Clouds follow wind patterns and deliver the water as rain on higher grounds. The water is then driven by Earth gravitation into lakes, dams and rivers. Hydropower as an energy source thus originate from solar and wind energy and represents, when the water has been trapped in a high altitude lakes, stored solar energy.

4.4.1 Basic Processes

Water stored at a height h above a power plant has potential energy. As the water falls, this potential energy is converted into kinetic energy until it reaches the plant. Neglecting friction, equating potential energy to kinetic energy we have:

$$m \cdot g \cdot h = 1/2 \cdot m \cdot v^2 \tag{4.14}$$

which gives the water velocity:

$$v = \sqrt{2 \cdot g \cdot h} \tag{4.15}$$

The mass flow of water passing the turbine per unit of time is (cf. the preceeding section on wind energy):

$$\frac{m}{t} = \rho \cdot A \cdot v \tag{4.16}$$

Using these two relations we obtain the power production formula:

$$P = \frac{mgh}{t} = g \cdot h \cdot \rho \cdot A \cdot v = \sqrt{2} \cdot \rho \cdot A \cdot (g \cdot h)^{3/2} \tag{4.17}$$

It is then understandable why hydropower is a very attractive energy source for countries with large h's (mountains and valleys). Compared to wind energy, the density of water is about 1,000 times larger than that of air. Thus, with $h = 100$ m, a 1 m^2 tube will have a theoretical loss free production equal to about 44 MW! Then what about efficiencies? The friction in the falling tube area can be made very small, around 10 %, thus bringing the remaining 90 % of the potential water energy down to the turbine. The turbine construction can also be made very effective, with around 90 % of the kinetic energy of the water being turned into rotational energy at the turbine. Then again, the conversion from rotational to electric energy is also in the 90 % region. The total effiency factor becomes the product of these three terms, ie. $\epsilon_w \sim 0.9^3 = 0.73$ or in many cases even higher. The reason for this high efficiency factor is the possibility to extract a much larger fraction of the water energy as compared to the wind energy restrictions in Betz law. In addition, concentrated water falling into a turbine can be handled in a much more controlled fashion than natural wind. The exact power production formula can then be written:

$$P = \epsilon_w \cdot \sqrt{2} \cdot \rho \cdot A \cdot (g \cdot h)^{3/2} \tag{4.18}$$

With enough supply of water, the power production can be held constant. However, in large scale production systems, the supply is seasonal, which again leads to seasonal power production.

There are three different hydroelectric systems (see Fig. 4.20):

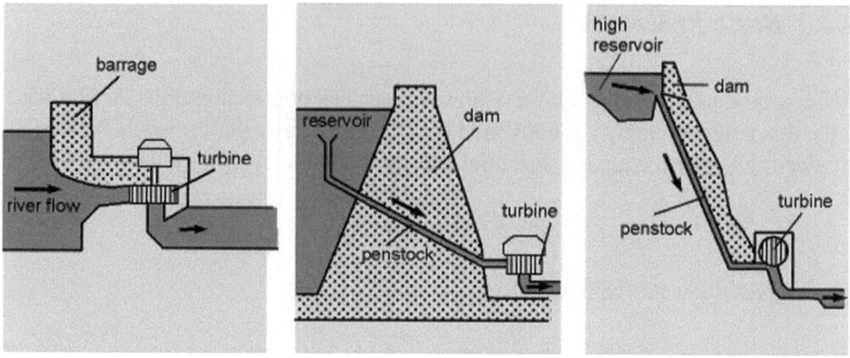

Fig. 4.20 From *left* to *right* run of river, dam system and high lying reservoir

- A dam system in connection to a river (run of river). The water gains much of its velocity as part of a river system and the inlet to the turbine region is directly along the water flow.
- A dam system which stores a high column of water. The inlet to the turbine is placed such that it maximizes h.
- A high lying reservoir which is connected to a low lying turbine creating a large h.

4.4.2 Present Use, Resource Considerations and Forecast

Hydropower's technical potential is estimated based on a region's topography and precipitation. Given that hydroelectric technologies are mature and that they have been cost-effective energy sources for decades, it is not surprising that a big share of the hydroelectric potential in many regions has now been exploited (see Table 4.4). Some insecurity remains with respect to the fraction of small scale production ($<$10 MW plants), which sites are generally not mapped in detail. In addition, the amount which can theoretically be put in production also depends on which price one sets as maximum acceptable.

With a total installed capacity of 990 GW at the end of 2012, the present world exploitation in terms of primary power production amounts to 3,700 TWh or about 0.4 TW. The biggest hydropower producers are, by order of importance, China, Brazil, Canada, the United States, Russia and Norway (BP 2013). Many developed countries (e.g. Austria, Canada, Iceland, New Zealand, Norway, Switzerland, South Korea) and developing countries (e.g. Albania, Brazil, Cameroon, Colombia, Costa Rica, Ethiopia, Ghana, Mozambique) cover at least 50 % of their electricity needs using hydropower.

From the three types of hydropower facilities considered in this section, hydropower combined with a high lying reservoir is the most attractive. It can contribute to the base load and also as balancing power since this type of plant can be

Table 4.4 The first column indicates the technical feasible hydroelectric potential in various regions (Bartle 2002)

Region	Technical feasible potential (TWh/yr)	Electricity production in 2012 (TWh)	Degree of exploitation (%)
Asia	6,800	1,504	22
South America	2,665	732.4	28
Africa	1,750	106.7	6
N. and C. America	1,660	690.8	42
Europe	1,225	598.5	49
Australasia	270	40.8	15
World	14,370	3,673.2	26

The 2012 hydroelectric production for these regions is displayed in the second column (BP 2013). The degree of exploitation of the technical potential is presented in the final column

ramped-up and down rapidly. The fast response characteristic of such energy type is valuable as it helps match the fluctuations between electricity demand and electricity supply. Run-of-river characteristics can be compared to those of wind and solar power as electricity will be produced whenever possible. Therefore, run-of-river hydro only contributes to the base-load as it is not dispatchable.

Hydropower is often considered by the public as being a green source of energy, whereas it is considered by certain scientists and communities as having a damaging impact on local and global environments. In evaluating hydropower, one has to look at a project scale as there might be an environmental cost and/or a societal cost attached to each project. The famous Three Gorges dam in China, the largest hydropower facility in the world with a capacity of 22.5 GW, forced more than a million people to move out of their villages as the plains they used to live on have been flooded, implying that the societal cost of this particular project is not negligible. Other impacts such as the possibility of controlling potentially deadly floods are positive.

In other cases, the environmental cost can be high. If a rainforest is flooded during the creation of a reservoir, the biomass will decompose and release tonnes of methane into the atmosphere and therefore contribute to the increase of the greenhouse gas levels in the atmosphere. In such cases, the emissions due to the project must be accounted for when producing electricity. However, large hydropower can be environmentally friendly if the environmental and social impacts of the project are considered thoroughly and avoided or compensated for.

Based on these considerations and on the potential presented in scientific studies (Bartle 2002), it is reasonable to estimate that hydroelectric power can grow to reach 1–1.5 TW on a global scale.

4.4.3 Cost

4.4.3.1 Resource Cost for Hydropower

Hydropower is dependent on water to produce electricity. Water is free but is however subject to a big opportunity cost, especially for hydropower plants using reservoirs as a storage facility. Managers of hydropower plants with reservoirs have the difficult task to maximize the plant's profit given a reservoir's capacity constraint and an uncertain water inflow. In a deterministic setting, where prices and water inflow are known, a plant manager would maximize its profit by using the water stored to produce electricity during high prices periods. Yet, in a realistic stochastic setting, general water inflow patterns are known but precise inflows are not predictable with certainty as they depend largely on weather and we all know how weather can change from year to year. By running water through the turbines now, a plant manager takes the risk of not being able to benefit from better market conditions in the future, especially if water inflows are inferior to predictions. By waiting for better market conditions, a plant manager takes the risk of losing profit, especially if water inflows exceed the predictions and storage capacity of the reservoir.

For smaller hydropower installations such as run-of-river, water inflows can be compared to wind and solar conditions. Electricity will be produced when there is enough water and when the water flow is sufficient.

4.4.3.2 Basic Cost of Hydropower

Most of the capital cost of hydropower with reservoir is related to the construction of the dam. The location of the project will therefore be a key factor in determining the profitability of a project. For instance, a location where little work needs to be done in order to create the reservoir will be more favorable than a location where building the reservoir is challenging and costly. Overnight costs of large hydropower installations is between 800 Euro/kW and 2,500 Euro/kW. For small hydropower installations (reservoir smaller than 30 MW and run-of-river), the cost is higher and if the overnight costs of some projects turns around 1,500 Euro/kW, the capital costs of other projects can exceed 8,000 Euro/kW.

It is to be noted that the expected lifetime of a large hydropower installation exceeds the expected lifetime of the other energy sources. The expected lifetime of a hydropower installation is 80 years and many examples of large hydropower dams over 100 years of age attest that the lifetime can be even longer.

O&M costs for both types are generally low since very little maintenance is required (Edenhofer et al. 2011). They are however substiantially higher for small hydro (3–25 Eurocents/kWh) compared to large hydro (1–10 Eurocents/kWh).

4.4.3.3 Electricity Generation Cost of Hydropower

The levelized cost of large hydropower installation is between 1.9 Eurocent/kWh and 7.7 Eurocent/kWh and the levelized cost of small hydropower installation is between

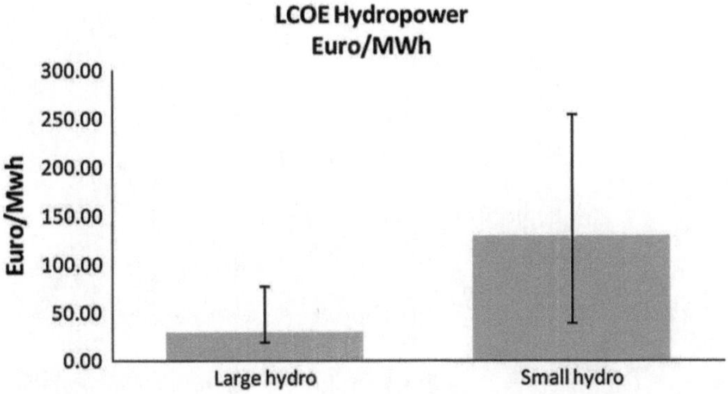

Fig. 4.21 Large and small hydropower plants LCOE

Table 4.5 Levelized cost of electricity for large and small hydropower installation

Label	Large hydro	Small hydro
LCOE (Euro/MWh)	19–77	38–255
Overnight cost (%)	87–92	60–98
O&M costs (%)	8–13	2–40
Fuel costs (%)	0	0

3.8 Eurocent/kWh and 25.5 Eurocent/kWh. The comparatively higher cost for small hydropower projects is partly due to O&M costs that can be two to three time as high compared to large hydropower installation, to a higher initial cost/kW and to much lower capacity factors in some cases (Table 4.5, Fig. 4.21).

Large hydropower is a mature technology. It is therefore unlikely that the costs of this technology will decrease significantly in the near future, even though several processes show potential for technological progress. Moreover, since hydropower costs are dependent to such a large extent on the location, it can be expected that capital costs for new plants will increase in developed countries due to the fact that the most favorable locations have already been developed. Thus, future development of hydropower in developed countries is mostly expected to be under the form of old facilities upgrading and in the run-of-river form, which is less mature and comparetively more expensive.

When it comes to developing countries, their hydroelectric economical potential largely remains untapped. Therefore, it is likely that most of the development in the future will take place in these countries, especially because hydropower is often a cheap, efficient and storable source of energy.

Fig. 4.22 A forestry swing machine at work in Kaibab National Forest, Arizona, USA (Credit: U.S. Forest Service, Southwestern Region, Kaibab National Forest)

4.5 Biomass

Biomass for energy production is probably the oldest energy source used by humans. Learning how to make fire with wood was one of the first and most important steps separating us from other mammals. And since then, the transformation of energy stored in wood has been used for heating and cooking all over the world. This concept of burning wood (and other fuels) to release the energy necessary for heating and cooking purposes is known as traditional biomass and it is still generally in use in low-income economies. The more modern approach of storing energy into a fuel and of transforming biomass into heat and electricity is refered to as modern biomass.

4.5.1 Basic Processes

Compared to other energy sources, biomass is complex. A multitude of materials can be transformed into power, heat or fuel. In the context of energy, biomass is defined as any vegetable matter that can be used to produce some kind of energy. Commonly, biomass is distinguished between solid biomass (straw, wood see Fig. 4.22), liquid biomass (vegetable oils) and gaseous biomass (biogas from decomposed food). Power plants burning biomass can produce electricity continuously and at a constant rate and thus serve in the baseload. Power plants burning biogas can produce electricity on demand and thus can serve in the baseload or as balancing power to

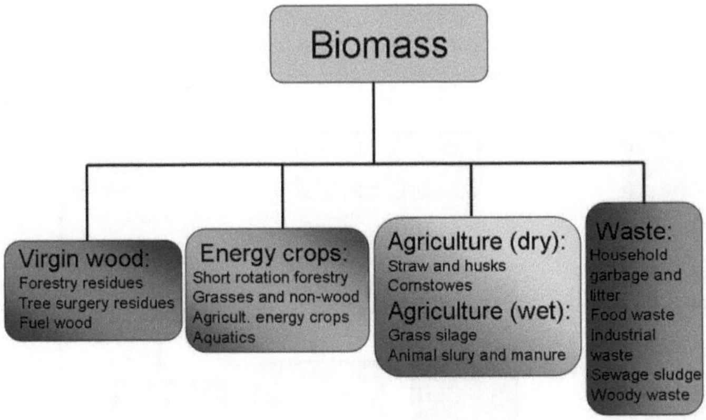

Fig. 4.23 Overview of sources for energy production based of biomass

help matching electricity demand and supply. Biomass can be used in the following general ways;

- To grow optimized plant materials for direct fuel for power plants for electricity and or heat production
- To grow plants for conversion into biofuels ("biodiesel")
- To use by-products and waste (second order biofuel) from agriculture, forestry, industrial processes and animal food.

At a more detailed level, the variety of fuel sources are much more complex, see Fig. 4.23.

The processing into useful energy forms then also follows a rather complex map as shown in Fig. 4.24. The main three processes are:

- Combustion: The biomass is here used as fuel in a regular power plant which creates electricity (and heat). The lower calorific value of biomaterial as compared to fossil fuels leads to an even higher fuel demand and consumption than fossil power plants.
- Dry chemical processes: The biomass is here processed either through pyrolysis or through gasification. Pyrolysis is an incomplete combustion process at around 250 °C where the biomaterial is heated without the presence of oxygen. Wood material is then transformed to charcoal which has a calorific value comparable to normal coal. The charcoal is then a raw materials which can be stored, effectively transported and used i.e. for heating when required. Gasification imply heating of the material to even higher temperatures which can transform the biomaterial into a range of gases or liquids, such as methanol. Again the end products can be used at later stages in specially designed energy transformation processes.
- Aqueous processes: The biomass is here heated to modest temperatures (250 °C) and mixed with suitable liquids and bacteria. The "hot biosoup" will then release

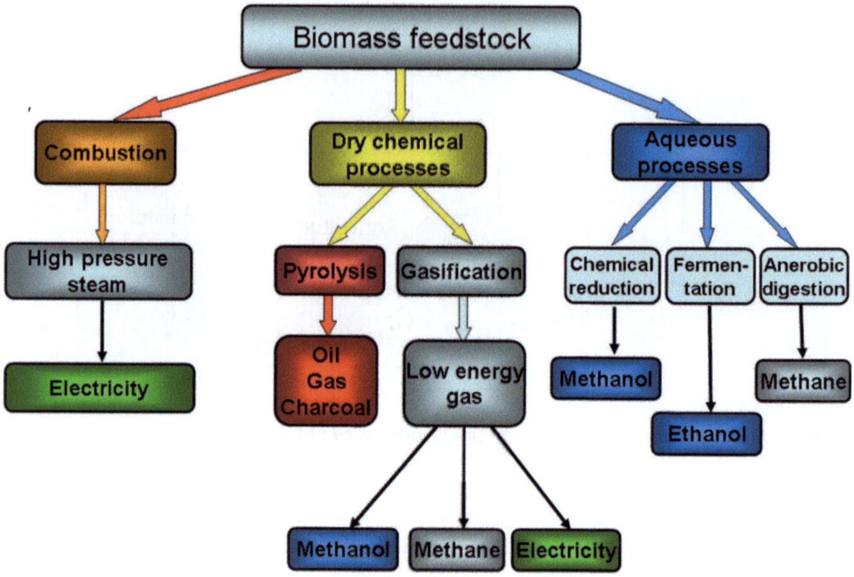

Fig. 4.24 Various ways of processing biomaterials

a range of useful liquids or gases based on chemical or anaerobic processes taking place in the soup. End products can be oils, methane and ethanol.

The end products of the first is electricity and heat while the end products of the last two are various oils and gas products. So, by setting up industrial processes based on these possibilities, we can create replacement for fossil fuels which today can still be extracted from the under ground in a cost-effective fashion. In addition, bioenergy production requires large areas which in a future world with more than 10 billion people will require correspondingly large areas for food production.

4.5.2 Potential and Area Requirements

As long as the use of plants for energy production is done at a rate which does not decrease the total amount of forest and plant areas, biomass is a truly a sustainable and renewable resource. From an environmental perspective, the use of biomass for energy production is then carbon neutral. In Fig. 4.2 it was pointed out that traditonal bioenergy currently accounts for nearly 10 %, which amounts to a power of around 1.5 TW. This makes biomass the fourth largest source of energy after oil, coal and natural gas. The large fraction of this number comes from direct heating and cooking around the planet.

How much additional power can be obtained from bioenergy is uncertain, because estimates regarding the resource potential of bioenergy show large discrepancies.

For example, a recent Swedish study Ladanai and Vinterbaeck (2009) estimates that bioenergy can have a resource potential of up to 30–40 TW, more than twice the present global power, while other studies conclude with 5–6 TW (Haberl et al. 2010).

So let us make our own rough estimate. In order to come up with such an estimate, the quantity of energy obtainable per area per time needs to be known, as well as some assumptions on on how much of the planet is actually available for bioenergy production.

The process which produces energy in plants is called *photosynthesis* and it is the driving life supporting mechanism in all plants, from trees and flowers to bacteria and algea. The essence of photosynthesis is the transformation of CO_2 extracted from the air to molecules containing carbon stored in the plant, in combination with the release of oxygen (O_2) back to the air. The photosynthesis reaction may be written:

$$6\,CO_2 + 6\,H_2O + \text{light energy} \rightarrow C_6H_{12}O_6 + 6\,O_2 \tag{4.19}$$

An average influx of light of 240 $W\,m^{-2}$ is utlized here to estimate the total bioenergy potential. About a quarter of the solar spectrum is applicable for driving the photosynthesis process and about 5 photons with energy ~2.5 eV are needed to store about 5 eV in a typical plant molecule. Thus, the efficiency of the photosynthesis process is about 35 %. There is however another major factor which reduces the energy absorption: basic molecules (CO_2 and H_2O) need to be available to plants at the same place where the solar photons are absorbed. CO_2 is instantly available in the surrounding air, but H_2O needs to be sucked up from the ground. This molecular transport takes time and energy and thus reduces the efficiency of at least by a factor 10. This translates to an expected power production of the order:

$$P_{bio} = 0.1 \cdot 0.25 \cdot 0.35 \cdot 240\ \text{W/m}^2 \sim 2\text{W/m}^2 \tag{4.20}$$

This estimate is a little optimistic when compared to real plant data (McKendry 2002). Typical numbers for the power density from wheat, poplar and mischantus are in the range 0.2–1.0 W/m^2 even though special fast growing plants, such as sugarcane at certain places may produce more than 2 W/m^2. However, by taking the combustion of the biomaterial into account, which have an efficiency in the range 30–40 % we arrive to a power density in full agreement with these data. Thus, when averaging over all growth and temperature zones of the planet, 0.5 W/m^2 seems a decent estimate of the power density.

The total global power production, assuming that the photosynthesis process takes place all over the planet then becomes:

$$P_{bio}^{total} = 0.5\text{W/m}^2 \cdot 4\pi R_{Earth}^2 \sim 250\ \text{TW}, \tag{4.21}$$

The real number is likely a factor 2 smaller since polar regions and deserts has much lower efficiency. Clearly, only a small fraction of this power can be used. Oceans

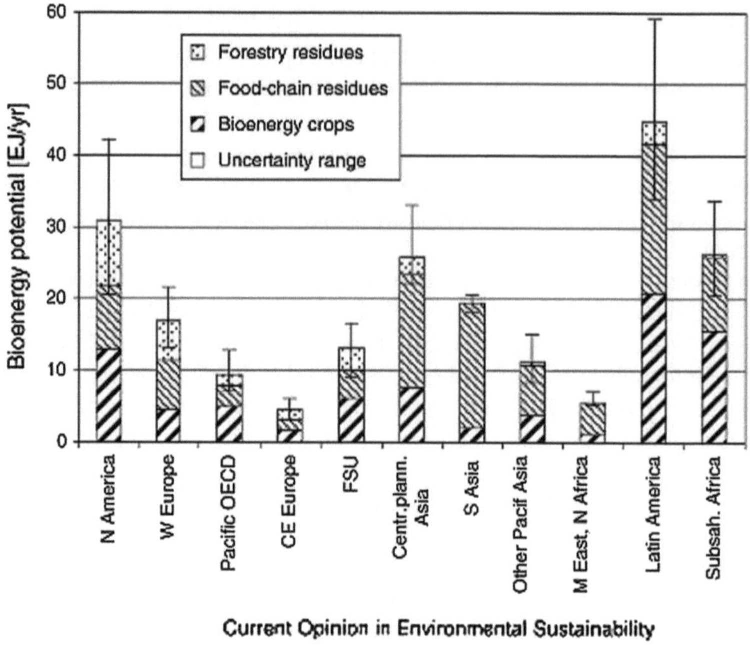

Fig. 4.25 Technical bioenergy-energy potentials in 2050, breakdown to 11 regions (Haberl et al. 2010)

cover about 70 % of the planet, leaving only 30 % for bioenergy production if we exclude bioenergy production in the sea. The total fraction of agricultural and forest covered land is about 60 %. Realistically a maxumum of 10 % of this area can be exploited for energy production, which leaves us with the global power potential of,

$$P_{bio}^{total} \longrightarrow 0.1 \cdot 0.3 \cdot 0.6 \cdot 250 \text{ TW} \sim 5 \text{ TW}. \qquad (4.22)$$

This number is then our estimation of the power absorbed from the incident solar spectrum which is available for biomass production.

It therefore seems rather optimistic to assume that the global potential for bioenergy is 30–40 TW (Ladanai and Vinterbaeck 2009), unless the totality of the forest areas and arable land is utilized for bioenergy production. An increase of our estimate could take place if new engineered plants can deliver more than 1 W/m^2, new technologies which utilizes waste products to a higher degree or harvesting bioenergy from the sea. It is shown in Fig. 4.25 how the total biopotential can be reached based on exploitation of forest residues, crops and various forms of waste in the various regions of the world. Taking these into account, we assume that the technical potential of bioenergy is around 6 TW.

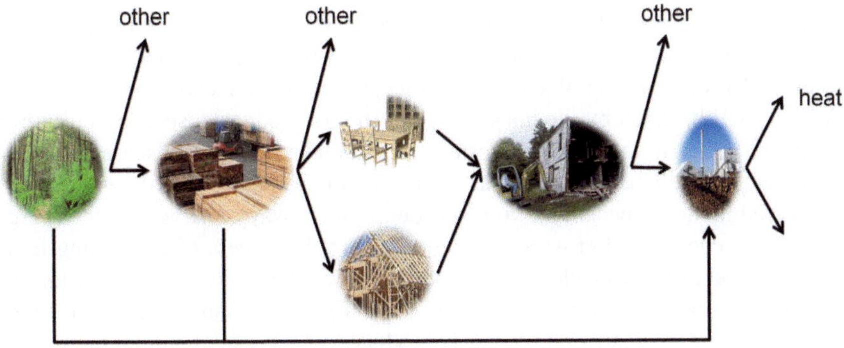

Fig. 4.26 A possible path for biomass power

4.5.3 Present Use, Resource Considerations and Forecast

Figure 4.26 was created with the aim of showing the reader that biomass is more complex than the other renewable energy sources. For example, in the case of wind, the energy from the wind is captured by the blades of a wind turbines and carried to the customer via the grid. With modern biomass, the path from the energy source to the customer is less direct. Let us consider the case of solid biomass, more precisely: wood. In a number of developed countries (Switzerland for example), it is not allowed to directly harvest wood from forests and burn it to generate electricity and heat. The possibility to create more value from wood if used in another industry combined to a political will to prevent unsustainable exploitation of the existing forests explain this interdiction. The link between wood and electricity and heat thus becomes indirect.

When a forest is exploited for economical reasons, the precious wood is sent to a saw mill where it is further processed. However, wood residues like branches and twigs are often left on site in the felled areas since there is little value attached to them. In some cases, some or all of these residues are collected and burned in a power plant to generate energy. These residues are the first, and most direct, possibility to use biomass for energy generating purposes. Once the wood is processed, it will be used to create furniture, as construction material or for other purposes. Residues will also be available, often in large quantities at this stage of the wood transformation process. It is the second possibility and still a relatively direct way of capturing some biomass for energy purposes. After the furniture and the constructions are demolished, the wood loses most of its value and is most often transformed in chipboards or valued in a biomass power plant, where the wood residues are burned after all undesired toxic materials (e.g. paint) have been removed.

Wood is not the only possible biomass resource as residues from food, food and plant processing and residues from animals can be gathered in compost. As these residues decompose, biogas, which can eventually be transformed into electricity, heat or fuel for transport, will be produced. Most supply chains are obvious but some can be more subtle. Consider a village where each household has a small garden where a handfull of vegetables is produced, resulting in the production of tiny quantities of by-products that would be too costly to be collected by an energy producer. The conclusion is that in such garden, the main products (e.g. carrots and/or potatoes) will be eaten by the household and the by-products disposed of, often in household compost. In larger composts, it would make economical sense to harvest the energy available but it does not at a household level. Is this the end of the story? Most certainly not, especially if you consider a 'dirty' way of harvesting the energy from these vegetables. Let us now explore the whole process of consuming food. Once the food as been swallowed by a person, it will be degraded and transformed into energy useful to 'power' the person. The body being unable to totally degrade the food, a certain quantity of 'waste' is produced which will end up in the sewage system. At large water treatment facilities, it is fairly easy to collect vast quantities of that waste and transform it into biogas, fuel that can be used for heating and electricity generating purposes. With a mildly open mindset, you can now see how the vegetables of your garden can heat up your house and bring light into your household.

From an energy perspective, biomass is not only used to generate electricity and heat but also as biofuels which can substitute conventional fuels from fossil sources. For example, corn, sugar cane, algaes and others can be processed to produce such biofuels.

Solid biomass power plants have been built in over 50 countries and the share of energy from biomass is increasing worldwide. Solid biomass is traditionally burned to produce electricity. About 83 GW of biomass power capacity was built by 2012, alltogether producing nearly 350 TWh of electricity annually (REN21 2013). The leading countries in terms of solid biomass power capacity are the United States (15 GW), Brazil (9.6 GW), China (8 GW) and Germany (7.6 GW). Often, the production of power is coupled with heat production that can be delivered directly to heat-demanding industries or to households via district heating systems.

Biomass is not the most well known source of energy but its contribution to the global renewable energy capacity is not neglictible. In some countries, the share of energy from biomass exceeds the share of energy from oil as it is the case in Sweden (32% vs. 31%) since 2009 (REN21 2011). Power from biomass is also becoming attractive in developing countries, especially because coal and natural gas-fired power plants can be retrofitted to enable co-firing with biomass.

Biomass can be converted directly into liquid fuels for transportation. Ethanol and biodiesel are the two forms of fuel that can be produced from biomass. Ethanol can replace or complement gasoline and help reduce greenhouse gas emissions. The quantity that can be blended (mixed with conventional gasoline) depends on the kind of vehicle, with modern vehicles increasingly being able to run entirely on ethanol. Biodiesel can similarily replace or complement diesel.

The biggest biofuel producers are the United States with a production of 50 billion liters of ethanol from corn in 2012, followed by Brazil with 21.6 billion liters of ethanol produced that year from sugar cane. Most major biodiesel producing countries are located in Europe, with Germany in the lead with an annual production of 2.7 billion liters in 2012. The global production amounted to 83.1 billion liters of

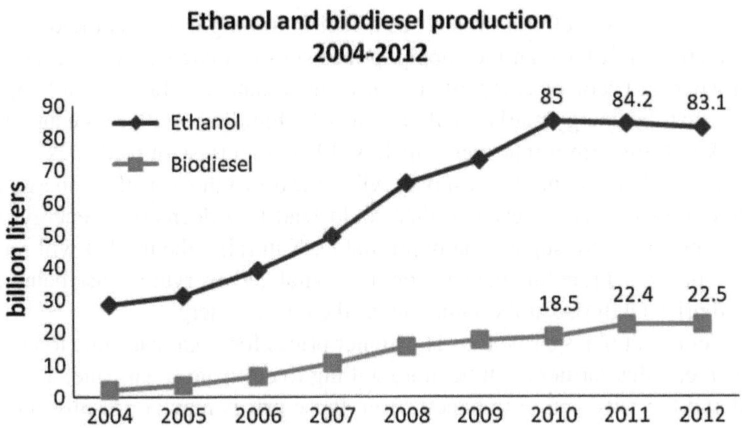

Fig. 4.27 Global ethanol and biodiesel production (REN21 2013)

ethanol and 22.5 billion liters of biodiesel (REN21 2013). Together, bioethanol and biodiesel contributed to 3 % of the global road transport fuels.

From Fig. 4.27, it is interesting to see that although global production of biofuels is generally increasing, the pace at which the production increases is slow. One of the reason behind this slow growth and even behind the drop in bioethanol production between 2011 and 2012 is due to a strong link between biofuel–food availability–food prices. This link will now be explored.

4.5.4 Crops for Food or for Fuel?

The main feedstock for biofuel production is corn in the USA, sugar cane in Brazil and cassava in Thailand. Besides the fact that these feedstocks are all first-generation biofuels,[5] they also have in common that they can be eaten by humans as well. Automatically, increased competition for a limited resource translates to higher prices. In reality, the link is more complex as it involves the price of energy, the cost of cultivating the feedstock and the cost of other feedstocks.

The supply curve of one commodity, corn for instance, is an increasing function; the higher the price, the larger quantity can be delivered (until a limit is reached). Other commodities, cassava or sugar cane, will also have increasing supply curves compared to corn. However, the quantity of a commodity delivered for a given price may differ between commodities.

Now, assume a country where sugar cane and other agricultural crops are cultivated. In addition, this country is endowed with some oil reserves. Initially, the

[5] The sugar or the vegetable oils found in these feedstocks can easily be extracted with existing technologies.

cost of using an agricultural product as a fuel is too high. Yet, as the demand for energy increases faster than the supply, prices start to increase. In fact, they reach a point where it becomes competitive to use sugar cane as a fuel. What happens in reality is that the energy market will take over (Schmidhuber 2008) the agricultural feedstock to transform it into fuel, which will lead to a raise in feedstock prices. In the energy market, the increased supply will slow down the growth in energy prices and increasing the supply even further could lead to a decrease in energy prices, thus making fuel from sugar cane unprofitable. Naturally, the market will reach an equilibrium. This direct link between prices is vital and everything else being equal, real food prices will rise at the same rate as the cost of energy.

Another effect plays a role too. The higher prices for sugar cane due to increased demand mean that farmers will be more willing to cultivate sugar cane. Yet, land is limited and the only option to benefit from these prices may be to substitute other agricultural crops for sugar cane. This will lead to a reduction in the supply of other agricultural crops, which in turns will drive up their prices. This indirect link is known as indirect price transmission through substitutes (Schmidhuber 2008).

The threat here is that increasing food prices will threaten people with a low-income as they may no longer be able to afford food. However, it also happens that these people are often employed in the agricultural sector, a sector which will benefit from higher crop prices. If markets are performing, higher crop prices will lead to a revitalization of the agricultural sector, leading to more employement and higher wages. The important point here is that increasing real food prices are a threat only if they exceed the increase in real income, which is likely in the short term.

The link between availability of food, price of food, bioenergy and energy price is complex and the aspects presented above are only some of the aspects affecting this link.[6] However, the story above can already be related to statistics. Figure 4.27 shows a decrease in the global production of bioethanol between 2011 and 2012. One of the reasons behind this drop can be explained as follows. Most of the ethanol is produced in the USA, which primarily uses corn as a feedstock. In 2012, a drought led to high corn prices, which in turn made corn less interesting for bioethanol production, so that the production of ethanol in the USA fell by 4 % that year (REN21 2013). This fall was enough to lead to a decrease in global ethanol output.

Now, other feedstocks, such as waste products, wheat straw, black liquor[7] and algae are also possible feedstocks for biofuel production. These feedstocks are known as second-generation biofuels, because the required fuel is harder to extract. The main benefit of these feedstocks is that they cannot be eaten, thus the question *for cars or for humans*, does not need to be answered, and they do not necessarily compete for land (algae production can be grown on land unsuitable for other crops). Moving to second-generation biofuels permanently, which might require further government intervention, could thus break the link between food prices and energy prices.

---

[6] For instance, government intervention in the biofuel market can render possible the fact that the price of biofuels rises faster than the price of energy.

[7] Waste from the paper industry.

4.5.5 Cost

4.5.5.1 Basic Cost of Biomass

The cost of biomass depends heavily on the type of technology and the purpose of the plant, the stage development of the process, the fuel used, the size and on the location. Data on solid biomass has been chosen to provide the reader with an idea on costs, even though costs for other technologies can be, and often are, very different. The use of solid biomass is however widely spread and focusing on that fuel is therefore relevant.

Capital costs of solid biomass amount to between 1,800 Euro/kW and 3,500 Euro/kW. Solid biomass-based power plants are less capital intensive than other renewable energy plants (e.g. solar, wind, hydro) since a big portion of the costs will occur during the life time of the plant. The capital costs thus only reach between 43 and 50 % of the levelized costs of electricity from biomass.

The importance of operation and maintenance costs can vary significantly. These costs have been estimated to account for between 3 and 35 % of the levelized costs of electricity. Operation costs can be significant for certain plants, especially if the handling processes and fuel pre-processing requirements are substantial. In some cases, the fuel is not obtained in a form appropriate for power production. It thus has to be transformed, dried and pre-processed (perhaps gaseified) and stored for some period of time before it can be burned. Small plants usually experience higher O&M costs compared to large plants.

4.5.5.2 Resource Cost for Biomass

The resource cost for biomass will depend on which resource is being used, on its price and on the location where the biomass is found. In countries where modern biomass has recently become popular, wood ships are becoming increasingly expensive whereas demolition wood can have a negative price as local community search to get rid of vast quantities of material. Even though some trade occures (and almost exclusively to Europe), local markets for biomass are to a large extent not connected to each other and large variations are seen in local prices. For example, the resource cost of solid biomass can be up to ten times more expensive in the Netherlands compared to the United States (OECD/Nuclear Energy Agency 2010) due to increasing demand and limited supply, varying land, labour costs and crop yields.

Fuel is an important part of the levelized costs since they account for 20–50 % of the final costs. If the feedstock use to produce power is a low-value by-product of another process, this share can be much lower (say around 10 % or lower), in which case the relative share of capital costs may approach 70 %. This range illustrates the difficulty in providing the reader with a definite answer on how much electricity from biomass exactly costs, especially since this range would have been even wider in the case other fuels would also have been taken into account.

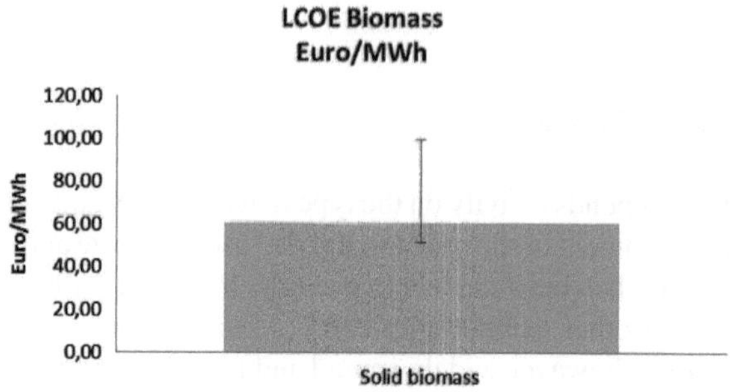

Fig. 4.28 Levelized costs of electricity from solid biomass

Table 4.6 Cost estimates of bioenergy (Edenhofer et al. 2011)

LCOE estimates	
Power generation (1–20 MW)	4.4–10.4 (Eurocent/kWh)
Hot water/heating/cooling (1–20 MW)	0.9–5.2 (Eurocent/kWh)
Ethanol (sugar cane, corn)	26–44 (Eurocent/L)
Ethanol (cassava, wheat)	52–70 (Eurocent/L)
Biodiesel (palm, jatropha)	35–69 (Eurocent/L)

4.5.5.3 Energy Generation Cost of Biomass

The levelized costs of electricity from solid biomass turn around 5.2–10 Eurocents/kWh, see Fig. 4.28.

Fuel costs and economies of scale play a big role in influencing the LCOE from biomass. In general, the LCOE of biomass for a specific technology is a linear function of the cost of fuel used (Edenhofer et al. 2011). The fuel costs of a power plant increase with its size since it is generally more challenging to organize more complex supply chains. However, the O&M costs decrease with the size of the plant (US Department of Energy 2011) at the same time, thus the size of the plant has to be studied carefully when designing a project.

Estimates (REN21 2011) for various outputs are provided in Table 4.6. The hypotheses and dataset used for these estimates differ from those used by the authors, thus the cost estimates obtained by the author and those presented thereafter should not directly be compared.

Research on alternative fuels associated with increasing experience and maturing technologies can hint that the cost of using biomass to generate power will decrease in the future, especially since it can lead to important greenhouse gases reductions compared to the use of more traditional fuels. Competition for feedstock, which drives up prices and the direct link between bioenergy and food security for some feedstocks (e.g. corn) may also prevent a significant deployment of some technologies. It is nevertheless not expected that the development of biomass will be uniform around the globe, especially since the biomass potential varies so significantly from country to country.

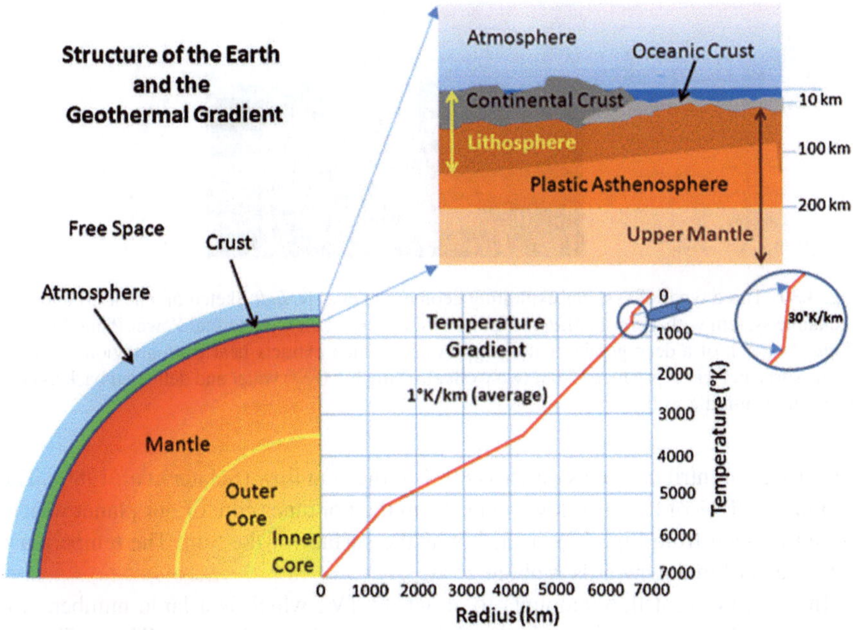

Fig. 4.29 Data showing temperature gradients and depth composition of the Earth. Reproduced with permission from http://www.mpoweruk.com/geothermal_energy.htm

4.6 Other Renewables

Four renewable energy sources are described in this section. These have in common that they have not evolved into commercial systems on a scale comparable to the previous four (solar, wind, hydro, biomass). Thus, cost estimates of potential future large plants cannot be made with reasonable certainty. However, it cannot be excluded that future technology developments of one or several of these technologies will lead to breakthroughs which may lead to large scale commercial deployment.

4.6.1 Geothermal

Geothermal energy relies on the heat of the Earth's crust to generates electricity. The total amount of geothermal energy is estimated as 3.5×10^{15} TWh. In other words, there is sufficient geothermal energy to satisfy the current global consumption of energy for the next 500,000 years, at least. However, to utilize this energy a significant transportation problem occur since the solid crust of the Earth is an effective heat shield which prevents the heat of the Earth center to escape, otherwise there would have completely different conditions for evolution of life on the planet. The radial flow of heat from the ground originates mainly from natural radioactive materials such as ^{238}U continuously decaying and releasing energy. Below a few meters depth,

Fig. 4.30 Three types of systems exploiting geothermal energy. *Left* Sketch of a near ground heat exchange system with a house. *Middle* Picture of the Nesjavellir Geothermal Power Plant, Iceland. *Right* A sketch of a deep geothermal energy system which extracts heat from a region 5–15 km below surface. Heat is produced from two wells showing hot (*red*) water and delivered back as cold water in the middle well

this process contributes to as much as 80 % to the heat flow (Sclater et al. 1980). The remaining 20 % of the heat flow comes from the hot inner core of our planet which, as can be seen from Fig. 4.29, is as hot as the surface of the Sun. The temperature increase per km downwards is about 25 °C.

In total, the heat flow amounts to about 32 TW, which is a large number. But divided by the surface area of the Earth it reduces to a modest 0.06 W/m^2. Thus, it is clear that geothermal energy at surface, on average, is an energy source with an extremely low intensity and which require special measures in order to be harvested. When that is possible, geothermal energy has the great advantage that the energy production is stable.

There are two existing and one possible future way of extracting geothermal energy:

- Reservoir for heat pump. By drilling a small distance downwards, say 100 m, the temperature is stable around 15 °C. If heat is extracted from this depth, it can be used as a reservoir for heat pumps and deliver heat to households. This can reduce the power consumption of a household significantly, on average about 30 %. In new advanced systems, such machines can also operate in reverse: to cool houses in the summer periods, store the heat in the ground and extract it during the winter.
- Near ground production.
  Several places on Earth are located near the border of tectonic plates. Here, the geothermal energy often manifests itself in the form of geysers and volcanos. In such areas, geothermal energy can be very cheap and is often explotied to a large extent. The middle panel of Fig. 4.30 shows the Nesjavellir Power plant on Iceland, which is an example of a power plant using near surface geothermal energy. This plant produces about 120 MW of electrical power per year and delivers around 1,800 L of hot water per second, servicing the hot water needs of the Greater Reykjavik Area.
- Enhanced Geothermal Systems (EGS). In these systems the idea is to drill several km downwards and extract the heat as hot water from say 10 km depth, where the temperature is around 250–300 °C. The very hot water produced can be turned

into steam and run turbines. No such plants have so far proven economically competitive since the drilling process is very expensive. An open question is how long and at which rate heat can be sustainable produced from such system. If the heat is extracted from dry rocks, the process will gradually suck energy from the rock and reduce the output of the plant. On the other hand, if the deep side of the plant is in contact with a wet, channeled or porus rock type, an aquifier or water reservoir, there may be sufficient heat exchange with a large deep ground volume which can make a plant economically productive for much longer time periods.

In the future, EGS systems may be established which may significantly contribute to increasing the contribution of geothermal energy to the world primary energy supply. Only in this case will geothermal energy make a leap upwards in its share of the total energy mix. At present, it is being utilized in some areas of the world such as Iceland and New Zeland as these countries enjoy optimal geothermal conditions.

More greenhouse gases emissions are released during the electricity production process from geothermal power compared to wind, solar or hydro. Yet, these emissions amount to less than 10 % of the greenhouse gas released during the process of generating electricity from coal for the same output. The low level of emissions is the reason why geothermal is often considered to be a clean source of energy.

The cost of geothermal energy depends mostly on three aspects, which are the relationship between temperature and depth, the reservoir rock's permeability and porosity and on the amount of fluid saturation (IDL 2006). The deeper the level of temperature necessary to generate power, the costlier the levelized cost of geothermal energy gets. High levels of rock permeability and porosity are necessary as it ensures good fluid flows in the geothermal reservoir. Lastly, a sufficient amount of fluid in the reservoir is necessary in order to ensure that the heat from the Earth can be captured. There is no fuel cost attached to geothermal energy as the natural Earth heat is free. The co-generation of heat and power usually improves the economics of a power plant.

In ideal locations, the levelized cost of geothermal energy lies between 3 and 5 Eurocent/kWh. In less favorable locations, the levelized cost of geothermal energy can reach three to four times these amounts.

4.6.2 Tides

The constant transition between low and high tides has its origin in the gravitational interaction between the Moon and the Earth, and between the Earth and the Sun. Since the Earth is not a solid spherical material, but rather contains mostly sea water at the surface and liquid rock in its interior, the gravitational forces create a drag towards the Moon and Sun which results in tides. If the water on the Earth had been uniform all over the surface, the tidal waves would have been almost periodic with 24 h cycles resulting in two high tide at two places of the Earth simultaneously: At locations closest to the Moon and at locations farthest away from the moon. Two

Fig. 4.31 Two types of tidal energy systems; tidal stream generators (*left*) and tidal barrage (*right*)

peaks and two lows would then be experienced within 24 h. Since the moon needs 28 days circling around the Earth, the largest tidal difference occurs when the Sun and the Moon are pulling in the same (or opposite) directions. Due to the distribution of large land areas, the tidal period and the difference between high and low tide vary quite a lot around the world.

The tides imply dissipation of energy from the Earth rotation to low quality random energy on Earth and to slightly larger distance (about 14 cm) between the Earth and the Moon every year. This causes the rotational energy of the Earth to decrease and create slightly longer days (and nights). The total power delivered to the tide is about 3 TW. Taking out 3 TW of rotational energy from the Earth each year or approximately 9.5×10^{19} J is very little, relatively speaking, since the rotational energy of the Earth is more than 2×10^{29} J. In other words, extracting 3 TW of the rotational energy of our planet would be theoretically possible for about two billion years. The intensity of the tidal power vary strongly from place to place. At certain good locations it may add up to 3 W/m^2 which is comparable to the energy intensity of wind energy. However, it seems extremely unlikely that more than a few percent of the tidal forces can possibly be harvested at a global scale, which limits the potential to a maximum of 0.1–0.3 TW.

There are essentially two technologies which can be used to extract energy from the tide, see Fig. 4.31. First, the water can be stored in artificial tide pools during high tide and when the tide level is lower outside the pool, this height difference then allows for energy extraction almost as a hydroelectric plant, Fig. 4.31 right. The second alternative is subsea versions of wind turbines; tidal stream generators. These are driven by tidal currents. Other more exotic concepts such as "tidal sails" exists as well.

An advantage with tidal energy is its stability. In contrast to wind energy for example, the energy production from a tidal energy system can be forecasted with precision as long as the equipment is operational.

Even though tidal power is seen by many as one of the new renewable energy source that will contribute to mitigating climate change and ensure energy security in the future, the concept of using tides to generate electricity is not new. For example, a 240 MW tidal barrage (called "La Rance") was installed in France already in 1966. Since then, one major installation has been commissioned in South Korea (Sihwa Lake tidal power station with its 254 MW) and a few other minor tidal installations have been built in Norway, Scotland, China and in the UK, bringing the total installed capacity to around 0.5 GW by end of 2012. This seemingly low interest for tidal power, (because not much is happening in the field for now, is due to the infancy of the technology (REN21 2013) and in the case of tidal barrage, to the very long recovery period associated to the initial investment.

The cost of tidal power depends mostly on the technology chosen and on the tidal range. Yet, all tidal power technologies are capital-intensive as no fuel cost is attached to tides. Generally, the tidal stream generators will have a higher levelized cost of energy compared to tidal barrage, due to a shorter lifetime, higher O&M costs and a less mature technology. Yet, the capital cost per kW of installed capacity is much lower for the former technology (tidal stream generators) compared to the second technology (tidal barrage) as no dam needs to be built to produce energy. In addition, the potential of tidal stream generator is many times higher than that of tidal barrage, which might justify further investment in the tidal stream generator technology.

Tidal power is a carbon-free technology, which means that it has the potential to help in the fight against climate change. However, environmental concerns may seriously slow down the deployment of tidal technologies, especially in the case of tidal barrage. A close monitoring of the ecological impact of the tidal power plant of La Rance showed that marine fauna and flora disappeared after the construction of the barrage due to the closing of the Rance Basin, as it led to the accumulation of organic matter in the basin and to salinity fluctuations. However, 10 years after the completion of the project, a new biological equilibrium had been reached in the basin. The Severn estuary case in Wales, UK is a sticking illustrative example of how environmental concerns may prevent the construction of a tidal barrage. In 2013, a completely privately funded 6.5 GW tidal power barrage across the mouth of the Severn river was proposed. If built, this project had the potential of meeting about 5 % of the UK's electricity needs. However, serious environmental concerns from adverse effects generated by this barrage, combined with a seemingly weak economic case, has led the government not to deliver the authorization to build the plant. This new failure to develop a tidal barrage in this estuary adds up to the many projects which have been rejected since 1920. Projects in other areas have also been rejected on the ground of their potential adverse environmental impact.

4.6.3 Waves

The energy stored and transported in waves is tertiary solar energy: A fraction of the solar energy is turned into wind and a fraction of the wind energy is transferred into energy carried by the wave. In general, both wave height and speed of the wave depend on the driving wind speed. A wave is formed from micro ripples in the water which are hit by the wind field. The oscillatory motion increases in height offering

Fig. 4.32 *Left* A large wavefront breaking towards the shore. The *middle* and *right panel* shows two possible technologies to transfer wave energy into electricity: The *middle panel* is the Pelamis wave energy collector which consists of four flexing sections which transfer energy by hydraulic generators. The *right panel* shows the AquaBuoy 2.0 concept which is lifted up and down by the waves. This motion can cause a piston to move which again can drive a turbine producing electricity

gradually more and more wave surface vertical to the wind field which again offers increasingly more effective transfer of wind energy into wave energy. The wave will grow in height and speed until an equilibrium height is reached when the wave travels with the same velocity as the wind. Then, the wave may propagate almost freely until it dissipates energy at the shore or at other obstacles. The idea behind wave energy is to harness the energy of the wave before it is lost by random dissipation.

The first attempt to harness energy from waves may have been as early as the end of the 19th century. Two of the existing prototype technologies are illustrates in Fig. 4.32. These all rely on a common principle: The energy of the wave is transferred into mechanical oscillatory energy of the system. The motion can then in the end drive a turbine system which finally produces electricity. Even though the fact that wave power can be captured to generate electricity or used for water desalination purposes has been known for decades, technical challenges and comparatively prohibitive economics have hindered the development of this energy source and only a few MW of wave power capacity is in operation today. Yet, several projects are planned in different countries around the planet, including Sweden, Australia, the UK and the US which, if built, can lead to a decrease of nowadays cost as gained experience may lead to better and more reliable designs.

Let us consider the physics of wave energy in some detail: Fig. 4.33 shows a schematic instant thin section of a water wave propagating to the right with velocity v, wavelength λ and average wave height h. The period, or time, between two successful wavepeaks at one place is T. Then the power of the wave equals the potential energy per period (c.f. Chap. 1):

$$P_{pot} = P_{wave} = \frac{m^* gh}{T} \tag{4.23}$$

where m^* is the mass per unit length (perpendicular to the wave) of elevated water approximated by the enhanced blues area in Fig. 4.33. Furthermore T is the period of the wave, i.e. the time it takes for the wave to undergo a full oscillation. During

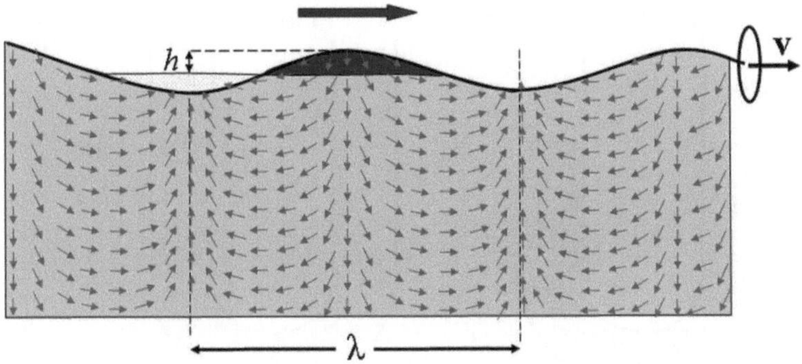

Fig. 4.33 Snapshot of an idealized water wave with wavelength, λ , velocity v and wave height, h. The wave propagates to the right perpendicular to the length of the shore

this time a point on the wavefront has moved one wavelength, so the wave velocity is $v = \lambda/T$. The mass can be expressed in terms of the wavelength λ, the water density ρ and a given unit length L perpendicular to the wave velocity. If we approximate the cross section of the elevated water with a triangle, the mass can be expressed as:

$$m^* = \frac{1}{2}\rho\lambda\frac{h}{2}L \tag{4.24}$$

Plugging this expression in to the equation above we obtain the formula,

$$\frac{P_{wave}}{L} = \epsilon_{wave}\frac{1}{4}\rho g h^2 v \tag{4.25}$$

which turns out to be approximately correct for deep water waves.[8] Here an efficiency factor ϵ_{wave} has been introduced. Plugging in the density of water, $\rho \approx 10^3$ kg/m^3, a wave velocity of 16 m/s , which is close to what has been measured by the Atlantic waves and $h = 1$ m, a rough estimate of one meter average wave heights, we obtain a power density of about 40 kW/m. The energy of more calm seas, like the Mediterranean is an order of magnitude smaller due to much lower average wind velocity. We estimate that the efficiency of electricity conversion can be $\epsilon_{wave} = 25\,\%$ which give 10 kW/m.

We are now in position to estimate the total potential of wave energy. First, it is realistic to assume that the wave energy can be harvested only once per length of coastline, because the energy which hit the coastline has been built up by 10 to 100 km of stable wind pressure. A device which extracts the energy outside the coast optimally would extract all the energy of the wave. Realistically, the efficiency is from 20–50 %. This implies that little is gained by narrow parallel lines of wave

---

[8] The derivation is in fact incorrect. Waves also carry kinetic energy, and in addition the relationship between (group) velocity and wavelength needs to be modified. In this particular case the modifications "cancel" and one is left with our present simple formula.

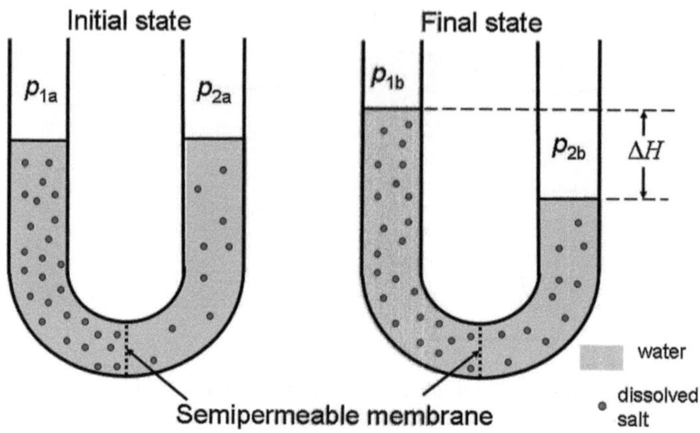

Fig. 4.34 Principle of osmosis

energy devices since the collision between the wave with the front row transforms energy to electricity and creates low quality random waves in addition. To extract wave energy twice per length coastline would need an order of 10 km separation between the line of wave energy transformers which would increase transmission and maintenance costs prohibitively compared to a single line of transformers.

The total world coastline is about 350,000 km which then gives an upper estimate of the technical potential of wave energy of 3.5 TW which is in agreement with literature estimates (Mork et al. 2010; Thorpe 1999) of about 2–3 TW. However, as in the case of biomass and wind it is completely unrealistic to envision the total world coastline covered by wave energy systems: Certain regions are too remote and others will have unfavorable wave conditions (small wave heights). If we assume that 10 % of the total world coastline is the maximum of what can realistically be used, we arrive at a real potential of global wave energy in the region of 0.3 TW.

The economics and final costs per kWh of wave power will depend on physical parameters such as wave speed, wavelength and water density and on the maintenance cost of extensive mechanical systems exposed to strongly varying forces over time. Yet, due to the little capacity installed to date and to the low level of maturity of the technology, any costs estimate would not be representative of the whole range of technologies being currently tested. Nonetheless, this cost is likely in order of magnitude higher than that of offshore wind power and solar energy.

4.6.4 Osmosis

Osmosis occur when salt water and fresh water are brought into contact through a semi permeable membrane. The fresh water molecules experience a force due to the difference in salt water concentration which cause a net water flux from the

Fig. 4.35 World's first osmosis power plant at Tofte, Norway. (*Credit* Bjoertvedt, Wikimedia Commons)

fresh water side to the salt water side. Equilibrium is achieved when the salt water concentration is the same on both sides and at that point a height difference between the two water levels is set up, as shown in Fig. 4.34.

The height difference, or equivalent, the pressure difference at the same height between the two water columns can be used to produce electric energy.

Osmosis energy plants can therefore in principle be built at any place in the world where rivers meet salt water or where either fresh water or salt water is artificially transported into contact with a sea or lake. The osmotic pressure between natural salt water and fresh water amounts to a height difference of 270 m. At especially salty waters, like the Dead Sea, the osmotic pressure amounts to a gigantic 5 km heigh difference. The global potential of the osmosis resource has been estimated to 2.6 TW (Nijmeijer and Metz 2010) which again imply that at maximum 0.2–0.3 TW is an upper realistic power utilization limit.

This source of energy is both new and very expensive and far from grid parity. However, research is made on various aspects of osmosis power and the aim is for osmosis power to reach a levelized cost of approximately 5–10 Eurocents/MWh by 2030. Also, an efficiency of 5 W of electricity per square meter of membrane is targeted. Osmosis also has an interesting feature in the sense that it can be used as a base load as the production is continuous in time. There exists one demonstration plant in Tofte in Norway since 2009 and the first full-scale osmotic power plant is planned by 2017 (Fig. 4.35).

In closing this section, it can concluded that the resource potential of osmosis, waves and tides are all rather small compared to the potential of sun, wind, bio and hydropower. An exception is geothermal energy which has a infinite potential but faces—at present—technical and economical large barriers which prohibit the energy source to be taken into commercial use on a grand scale.

4.7 Storage and Energy Carriers

It must be clear by now that some renewable electricity generating technologies are intermittent sources of power. In addition, optimal renewable energy sources are often located in remote areas. These elements point to the need for storage to suppress the intermittency issue as well as the need for energy carriers to bring the energy from remote areas to consumption points.

Fossil fuels are transported to the power plant via the use of pipelines, trains or tank ships, where the energy can be transformed in useful energy. Wind and solar irradiation cannot be transported in similar ways. So, renewable energy can either be transported via electricity transmission lines, stored as chemical or potential energy or stored in an energy carrier. Storage includes batteries and pumping storage, whereas hydrogen, methanol, peroxide or anhydrous ammonia are example of energy carriers. Among those, only hydrogen and methanol will be discussed since the principles behind the other energy carriers are similar by nature, and also because hydrogen and methanol are being produced at a commercial scale.

4.7.1 High Voltage Direct Current Transmission Lines

One of the major breakthrough opening for the possibility to build large and efficient power plants was the invention of electricity transmission lines, because the electricity generated centrally could then be brought to consummers via electric cables. There are basically two types of transmission lines, the alternative current (AC) lines and the direct current (DC) lines. In an AC line, the flow of electric charges reverse direction periodically, whereas the flow is only in one direction in DC lines. Most households appliances rely on AC, whereas battery-powered appliances rely on DC.

AC lines have become popular because they rely on the principle of magnetic induction.[9] Thanks to this principle, it is cheap and simple to raise or lower the voltage of a transmission line. However, efficiency issues plague AC lines and electric losses can be significant, which implies that connecting remote power plants (farther than 800 km away) to consumption points can be senseless.

Best renewable resources (e.g. solar conditions are best in desert areas along the equator) are example of remote energy resources and it may not be possible to tap

9 Magnetic induction means that an electric current is induced by a time varying magnetic field.

into these resources if AC lines were the only possibility to distribute the electricity generated in remote locations. High voltage direct current (HVDC) transmission lines can help here. The efficiency issues of AC lines are tackled with DC lines and losses are significantly reduced and are of the order of 3 % per 1,000 km. These lines are only suited for transportation over long distances because the inverters needed to transform from AC to DC are expensive and lead to energy losses.

In a renewable energy society, it is likely that the backbone of the energy system would rely on HVDC lines to allow for the geo-spread and techno-spread to take effect in full.

Practical Example: DESERTEC

In 2003, the Trans-Mediterranean Renewable Energy Cooperation, which became the DESERTEC Foundation in the meantime, launched a project planning on connecting the renewable power plants (solar, wind, biomass, hydropower and geothermal energy) in Europe, Northern Africa and the Middle-East and to tap into the energy potential from the Sahara deserts in order to provide these regions with clean energy. HVDC lines are needed to make this project technically feasible (Fig. 4.36).

For example, HVDC lines could mitigate any intermittency issues due to the extensive use of wind power in Spain by bringing electricity from the Norwegian hydropower dams.

Despite indicating the scale of the projects needed to reach a more sustainable future, the Desertec concept also illustrates a potential threat caused by interconnecting many countries together. The solar energy wanted by renewable energy hungry EU countries in order to reach a carbon-free future means that Europe is likely to depend, to some extent, on countries located in Northern Africa in the future. The "Arab Spring", which started at the end of 2010 showed the risk for Europe to rely on other countries during times of political instability. The outcome of this wave of demonstrations and protests is likely to slow down the ardour of some European politicians in favor of such projects, at least in the medium term, as well as slowing down the enthusiasm of the companies involved in the project as they realise that the political risk of this massive project may have been undervalued.

4.7.2 Batteries

A battery is a mean of storing energy until it is needed. A chemical battery consists of one or several electrochemical cells which convert stored chemical energy into electrical energy. Each cell consists of a negative electrode, an electrolyte, a separator and a positive electrode. When a device needing energy is connected to a cell, a current

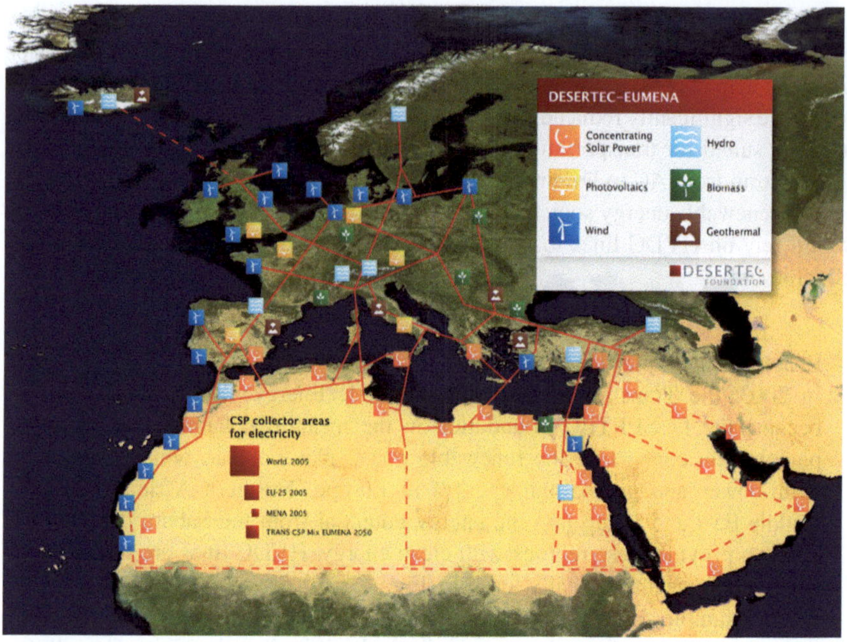

Fig. 4.36 The DESERTEC project. *Source* Desertec Foundation, 2011

of electrons flows from the negative electrode to the positive electrode through the device, providing the latter with electricity. The chemical reaction ceases as soon as the device is disconnected from the cell (Fig. 4.37).

It was already in the 1800s that Alessandro Volta invented the first battery (the Voltaic Pile) when he discovered a practical method of generating electricity. There now exist two categories of batteries. Primary batteries convert their chemicals into electricity only once and cannot be recharged. Secondary batteries can, on the other hand, be recharged because their electrodes can be reconstituted by passing electricity back through it. Secondary batteries are therefore of interest in applications where charging, discharging and recharging of the batteries are needed. Electric cars are one such application as is the use of batteries to mitigate the intermittent electricity generation from solar and wind power projects.

Technological progress has been made overtime and batteries have become lighter, cheaper and more efficient. However, progress still need to be made before that they can really support the renewable energy industry. Taking the example of electric cars, batteries are often qualified as being dangerous because they are made with hazardous substances, they are difficult to dispose of or to recycle, they take too long to recharge, they are heavy and they do not hold enough charge. These drawbacks limit the range of existing electric cars, making it challenging for them to significantly penetrate any market. The challenge is thus to make the batteries more powerful, lighter, smaller and especially, to reduce their costs. However, there is a steadily development in the

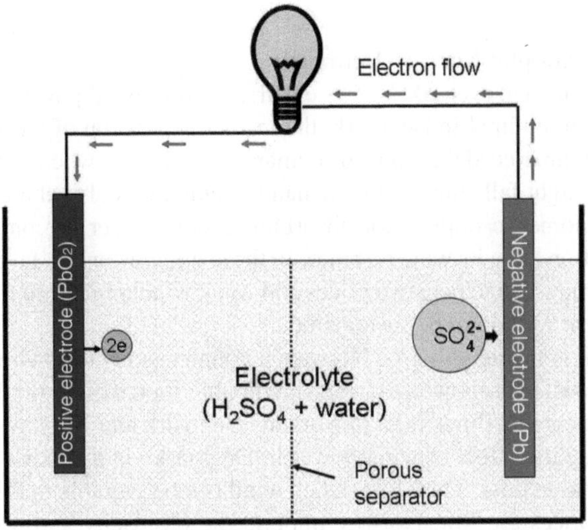

Fig. 4.37 How a lead-acid battery works. SO_4^{2-}-ions in the sulfuric acid transport electrons through the electrolyte so they can flow in an external circuit

auto industry to meet these challenges and new and improved models, both fully electric and hybrid cars, enter the market every year.

4.7.3 Pumping Storage Hydro

A storage possibility that works well in practice is the concept of pumped storage hydro. The idea behind this concept is fairly straightforward. Water is pumped into hydropower reservoirs when electricity prices are low. Pumping storage is therefore a solution to absorb excess energy generation. The energy is thus stored as potential energy, which can be reused later, when electricity prices are high.

Pumped storage hydro may also become more attractive as the share of electricity from intermittent technologies increases. At high shares of intermittent power, there can be time when supply exceeds demand, thus leading to negative pricing. Worse than negative prices is when demand exceeds supply, leading to blackouts. Pumped storage hydro can be used to value the extra energy when supply is high (and prices low or negative), while providing ancillary services to meet demand when energy supply is insufficient.

The energy penalty of using pumping storage is approximately 20–25 %, meaning that up to a quarter of the energy is 'lost' in the process.

138 GW of pumped storage hydro were operational at the end of 2012 and it is very likely that more capacity will be added in the future.

Practical Example: Norway-Denmark

Denmark generates over 20 % of its electricity from wind power. When wind conditions are optimal in Denmark, the instant generation of electricity from wind power can exceed the domestic demand. Conversely, when wind is scarce, the supply might fall short of the demand. Denmark is also characterized by its lack of domestic options for absorbing excess power or compensate for low production from its wind turbines. In these circumstances, favorable wind conditions may lead to negative prices and weak wind might lead to blackouts. This situation would not be sustainable.

Denmark is the neighbor of Norway, a country generating virtually all its electricity from hydropower, a cheap and flexible source of power. In addition, pumping storage facilities exist in Norway. Denmark and Norway collaborate on mitigating the effect of wind power in Denmark via the existence of sub-sea transmission lines. Therefore, when wind is too favorable in Denmark and electricity prices are low, Norway imports the electricity and uses it either directly or for pumping water in its pumping storage facilites. When wind is scarce in Denmark and electricity prices are high, Denmark can import electricity from Norway. This scheme therefore serves to balance the electricity market in Denmark.

It is statistically challenging to conclude how much of the electricity produced from wind power in Denmark eventually needs to be exported in order to cope with the high share of wind power. A recent study Lund et al. (2010) indicates that this number could be as high as 36.9 % for 2008, although it is likely to be lower.

4.7.4 Hydrogen

Hydrogen is a chemical element which can be used to store energy and serve as an energy carrier. Hydrogen in itself is the simplest and lightest chemical element and it is generally found in combination with other elements. Water, or H_2O, is a well known chemical substance combining two molecules of hydrogen and one of oxygen (11.2 % hydrogen by weight). The elements composing water can be split to extract molecular hydrogen H_2 by *electrolysis* where electric current is conducted through water via an anode and a cathode. The cathode reaction is:

$$2H_2O + 2e^- \rightarrow H_2 + 2OH^-$$

and the anode reaction:

$$2OH^- \rightarrow \frac{1}{2}O_2 + H_2O + 2e^-$$

The chemical reaction is endothermic, which means that an amount of energy (237 kJ/mol) is needed as input before the elements composing water can be split.

An other method for producing molecular hydrogen gas is starting from *natural gas* and supplying energy through the chemical reaction:

$$CH_4 + 2\,H_2O + 165\ \text{kJ/kg} \rightarrow CO_2 + 4\,H_2$$

It emerges from the process above that a certain quantity of carbon dioxide is released when natural gas (or any fossil fuel) is used to extract molecular hydrogen. This issue can be suppressed by using renewable power such as wind or solar power. Water molecules can then be split by electrolysis. As long as the molecular hydrogen is kept away from oxygen, the embodied potential energy is stored. This stored energy can then be transported to a demand point. When put in presence of air, the molecular hydrogen will react with the oxygen if sufficient energy is introduced to the mixture. An exothermic reaction occurs in the process, meaning that energy is released, and water is formed. The released energy can then be used to power a vehicle for instance. If wind power is used to split the water molecule, about 20 % of the energy is lost in the process of splitting and reforming water molecules. Nevertheless, associated to intermittent energy generating technologies, the use of hydrogen reduces or suppresses the need for back-up power since the energy becomes dispatchable.

The two main benefits of hydrogen are that energy can be stored more easily (with presently available technologies) than electricity and that hydrogen can facilitate the valorisation of remote energy sources. This means that the use of hydrogen can compete with high voltage direct current lines for this purpose.

Some drawbacks are associated to the use of hydrogen as an energy carrier, namely security and lack of existing infrastructure. Hydrogen has the highest energy to weight ratio of all fuels (Royal Belgian Academy Council of Applied Science 2006) and has an explosive range[10] greater than methane. In addition, hydrogen mixes with air considerably faster than methane. Finally, hydrogen is lighter than air. It means that in open air, released hydrogen will have a tendency to rapidly disperse and escape vertically (as opposed to concentrating on the ground as gasoline would do). In a constrained environment, hydrogen can however lead to destructive explosions.

The second issue is the lack of existing infrastructure. At present, up to 3 % of hydrogen by volume can be added to natural gas without requiring any change to existing devices. A widespread use of hydrogen is thus tributary to a new and extensive infrastructure for producing, distributing and storing the energy carrier. Devices (e.g. cars) dispensing the energy will also have to be reinvented which will require a very large amount of capital.

[10] The explosive range is the proportions of combustible vapor mixed with air which leads to an explosion if ignited.

Practical Example: Utsira, Combined Wind and Hydrogen
Utsira is an island 20 km off the coast of Norway. Over two hundred inhabitants live on Utsira. Norsk Hydro (which became part of Statoil) started a combined wind power—hydrogen project on the island in 2004. The idea behind the project was simple; when it was windy, the inhabitants of Utsira would use the electricity produced from the two 600 kW each wind turbines. Any extra energy was stored in the form of hydrogen. When the wind would stop, a hydrogen combustion engine totalling 55 kW of installed capacity would provide the people of Utsira with power, ensuring the reliability of a clean and sustainable energy system (Fig. 4.38).

4.7.5 Methanol

Another form of storage and energy carrier is methanol CH_3OH, which is the simplest alcohol. Several processes are used in practice to produce methanol, one of them being via natural gas. As in the case of hydrogen, the chemical process is endothermic. Methanol burns in the presence of oxygen with the following chemical reaction:

$$2CH_3OH + 3O_2 \rightarrow 2CO_2 + 4H_2O$$

Water and carbon dioxide are released in the process. Methanol can be a cleaner form of storage and energy carrier if renewable energy is used in the process of forming methanol molecules, in which case it is sometimes refered to as renewable methanol.

Renewable methanol is the result of an emission to liquid process, where carbon dioxide is captured and mixed with water through a system of electrolytic cracking and catalytic synthesis to produce methanol. The chemical reaction behind renewable methanol is:

$$CO_2 + 2H_2O \rightarrow CH_3OH + 1.5O_2$$

In order to be effective, the process needs to capture carbon dioxide released during a normal industrial process and have access to electricity from renewable sources. Renewable methanol is a source of energy storage that can later be blended with conventional fuel or fed directly to cars as a fuel in the transportation sector. The potential is limited only by the amount of carbon dioxide emitted annually and by the availability of renewable energy. The carbon dioxide needed in forming methanol is eventually released, contributing to increased concentrations levels of carbon dioxide into the atmosphere. Nevertheless, the process is considered carbon

Fig. 4.38 Picture of Utsira, the two wind turbines can be seen in the background (courtesy of Kurt Misje ©)

neutral since the carbon dioxide would have been released at an earlier stage without being transformed into methanol.

Practical Example: Emissions to Liquid, Iceland

Iceland has access to excess near-surface geothermal energy which goes to waste if untapped. Because the needs for electricity and heat in Iceland are already met, one of the few remaining possibilities to exploit geothermal further is in the transportation sector. However, electrification of that sector is costly and challenging. Iceland is also characterized by a lack of fossil fuel resources, and gasoline and diesel therefore need to be imported at great costs, either from the United States or from Europe.

A firm contributes to solving the fossil fuel issue in Iceland by transforming geothermal energy into methanol and thus further increase the value of Icelandic geothermal energy. Some carbon dioxide is released in the process of transforming geothermal energy into electricity and heat. The emission-to-liquid firm thus captures the carbon dioxide from the geothermal power plants and combines it with water to obtain methanol, which can be mixed with gasoline. In the long-run, this process may allow Iceland to export its geothermal energy to countries starving for clean energy.

The development of storage technologies will strongly influence the economy and the buildup of renewable technologies and favor the energy transformation technology which fits with the storage technology. Batteries are perhaps the most urgent storage system and batteries are currently undergoing constant improvements due to technology development. This is not (yet) the case for hydrogen and methanol production, simply because the quantity is cheaper to manufacture from natural gas. Other more exotic transmission systems has also been suggested, such as transmission of electromagnetic energy through open air by microwave fields and high temperature superconducting cables which allow for loss free transmission of electric energy. These are examples of processes at present known to work on the laboratory, while the economic of upscaling to commercial systems remains to be seen in the future.

4.8 Exercises

1. **Tide-Waves**

   (a) What is the physical origin of tidal power and what is the advantage of tidal power compared to wind power? (Please answer with max two sentences.)
   (b) Calculate the maximal power per meter in sea waves approaching a coast with velocity $v = 16$ m/s and height $h = 1$ m. Give also your estimate of the efficiency of a real wave energy system.
   Assume you require 0.5 GW from a wave or tide based power plant. Assume tidal energy can be extracted from a subsea "tidemill" system with rotor radius 3 m and average current velocity $v = 1$ m/s.
   (c) How long coast line is required for the wave based power plant and how many tidal mills are necessary for the tide based power plant?

2. **Wind**
   The power of a single wind turbine is proportional to the formula for $P = 1/2\rho A v^3$, where the density of air, $\rho = 10^{-3}$ kg/m^3, A is the area spanned by rotor blades and $v = 6$ m/s is the average wind speed. Assume that a rotor diameter is 75 m.

   (a) Calculate a maximal power in the wind hitting the mill and give an estimate of the expected power. Describe the most important mechanism(s) which modify the max effect.
   Assume a windfarm needs a windmill separation of 5 times the size of the rotor blade diameter.
   (b) Calculate the expected power per square meter of the wind farm and the area needed for a 2 GW wind farm.
   (c) Discuss the economic tradeoffs of building wind turbines at sea versus on land.

3. **Solar basics**
   The power from the sun is $P_{sun} = 3.85 \times 10^{26}$ W, the distance between Earth and the sun is 1.5×10^{11} m and the radius of Earth is 6.38×10^6 m.

(a) Calculate the solar power per area hitting the top of the Earth atmosphere. The average power of solar energy hitting the Earth ground is 240 W/m^2.
(b) Describe the most important processes leading to this result.
(c) Describe two technologies to produce energy from the solar radiation and give an estimate of how much power can be expected from each source.
(d) What is the origin of the greenhouse effect and which gases are the most important for the effect?

4. **Solar**

(a) Describe with max 1–3 sentences the physical principles behind photovoltaics.
(b) Explain why (max 1–3 sentences) the efficiency of a solar panel decrease with increasing cloud coverage.
(c) What are the economic advantages and disadvantages of the three major technological forms of solar power: Silicon wafer based, Thin film, and Concentrated solar?
(d) Why can some forms of solar power be considered economical at a higher electricity generation cost than other renewables such as wind or hydro?
(e) Estimate the area needed for a 1 GW thin-film solar panel system placed near Equator. Do the same exercise for a silicon-based solar panel system. (Hint: The total radiating power hitting Earth is 1.74×10^{17} W and the radius of our planet is 6.37×10^6 m).
The International Space Station (ISS) orbits the Earth at around 300 km altitude with orbital period $T = 90$ min. It is powered by 8 solar panels, each with extension up to 35×12 m^2.
(f) Assuming commercial solar cell efficiency, estimate how much electrical energy the station can receive in one "Earth day" (24 h).
You have been mandated to install a 1 GW system and you have the option between thin-film solar panels and silicon-based wafers. Thin-film panels come at a capital cost of 2,000 Euro/kW of installed capacity and the operation and maintenance costs amount to 4 % of the capital costs per year. These numbers are 3,000 Euro/kW and 1 % for silicon-based systems. The economic plant life of either type of panel is 25 years and a load factor of 22 % can be achieved. Assume a discount rate of 10 % and no escalation rate.
(g) How is it possible that the load factor is the same for both projects?
(h) Calculate the levelized costs of electricity for both types. Which one would you choose?
(i) Can you think of any circumstance under which you would choose the other technology?

5. **Solar bis**

You are renovating your house and you are hoping to cover your own electricity needs with sustainable electricity. A vendor approaches you with the following offer to put solar panels on your roof. The major elements of the offer are reproduced below.

- 42 panels of 240 W of peak capacity each will be installed. The total capacity amounts to 10,080 W.
- Due to the orientation of your roof, you lose 15 % on the average yearly electricity production.
- Average yearly production (15 % loss taken into account): 10,800 kWh.
- Cost for the panels: CHF 29,400.- (CHF means Swiss francs)
- Cost for the modules, balance of system and inverters: CHF 11,358.
- Labor costs: CHF 6,400.
- VAT 8%: CHF 3,780.

In addition, after a quick search, you discover that the economic plant life of such project is usually of 25 years and that you can expect O&M costs of 12. CHF/MWh.

(a) Using a discount rate of 10 %, calculate the LCOE of this project.
(b) You are informed about the possibility to sell the electricity you generate onto the grid at a price of 0.46 CHF/kWh, tariff valid for the 25 years of the project. If you compare this amount to the LCOE you have just estimated, will you be able to cover your costs over the lifetime of the panels?
(c) If you have the possibility to place the panels differently and therefore benefit from optimal solar conditions (the 15 % loss disappears): how does that affect your costs? Is the project profitable under the feed-in tariff?

6. **Bioenergy**
Our planet has a total surface area of about 5.1×10^8 km^2, of which about 70 % is covered by oceans. The world's surface land area contains about 13 % arable land and about 40 % of it is covered by forest. The average solar radiation per area is 240 W/m^2.

(a) Describe the most important physical factors which lead to an estimate of the bioenergy efficiency of less than 1 W/m^2.
(b) Estimate the total global potential of bioenergy based on the assumption that 10 % of all arable land and forests is exploited for power generation. Use a power intensity of 0.5 W/m^2
Assume that a power plant generating electricity is planned. We are interested in knowing what size/capacity the plant could be and also to get an idea of the cost of fuel. The LCOE of this plant (excluding fuel) amounts to 20 Euro/MWh. The fuel would come from a nearby large forest and the calorific value of the forest material is set to 15 MJ/kg in average. It is

estimated that 100 km^2 of the forest can be sustainably used as input material for the power plant each year.

(c) With a power intensity of 0.5 W/m^2 from the forest and 40 % efficiency of the conversion of heat to electricity, what is the maximum electric power production of the plant?

(d) Over each year of operation, the plant is expected to be under maintenance for 15 % of the time. With this information, what should be the installed capacity of the plant (round it to the nearest MW)?

(e) What is the mass of harvested wood material per square meter used for fuel? It costs 0.05 Euro to extract 1 kg of wood material and transform it into fuel for the plant. Assume a discount rate of 5 % and an escalation rate of 1 %.

(f) Find the necessary average electricity price to make the plant profitable over a 20 year perspective. (If you cannot solve question e), use 100 Mkg wood per year for fuel.)

7. **LCOE of Renewable Energy and Net Present Value**
Information on a tidal system is provided in the table below. Estimate this system's LCOE.

Label	Unit	Quantity
Net capacity	MW	304
Full load hours	Hours/year	2,628
Total overnight costs	USD	793,744,000
Fuel costs	USD/MWh	0
Yearly fixed and variable O&M costs	USD	149,796,000
Discount rate	%	5
Escalation rate	%	0.5
Economic plant life	years	75

(a) Feed-in tariffs for this type of system are set at 0.28 USD/kWh. Is the plant owner expecting to cover his costs over the lifetime of the plant?

(b) Using the net present value approach, how many years will it take you before the present value of the project becomes positive?

References

Archer, C.L., Jacobson, M.Z.: Evaluation of global wind power. J. Geophys. Res. **110**(D12), 2005. doi:10.1029/2004JD005462

Bartle, A.: Hydropower potential and development activities. Energ Policy **30**(14), 1231–1239 (2002). doi:10.1016/S0301-4215(02)00084-8

Betz, A.: Introduction to the theory of flow machines (Randall, D.G. Trans). Pergamon Press, Oxford (1966)

BP: Statistical Review of World Energy 2013. http://www.bp.com/content/dam/bp/pdf/statistical-review/statistical_review_of_world_energy_2013.pdf (2013)
Edenhofer, O., Pichs-Madruga, R., Sokona, Y., Seyboth, K., Matschoss, P., Kadner, S., Zwickel, T., Eickemeier, P., Hansen, G., Schlomer, S., von Stechow, C.: Special report on renewable energy sources and climate change mitigation. Intergovernmental Panel on Climate Change, Working group III—Mitigation of Climate Change (2011)
EPIA/ Greenpeace: Solar generation 6—solar photovoltaic electricity empowering the world 2011. European Photovoltaic Industry Association. Brussels, Belgium (2011)
Gustavson, M.R.: Limits to wind power utilization. Science **204**(4388), 13–17 (1979). doi:10.1126/science.204.4388.13
GWEC: Global wind report 2012. Global Wind Energy Council (2013)
Haberl, H., Tim, B., Bhattacharya, S.C., Erb, K.-H., Hoogwijk, M.: The global technical potential of bio-energy in 2050 considering sustainability constraints. Current Opin. Environ. Sustain. **2**(5–6), 394–403 (2010). doi:10.1016/j.cosust.2010.10.007
Hoogwijk, M., de Vries, B., Turkenburg, W.: Assessment of the global and regional geographical, technical and economic potential of onshore wind energy. Energ. Econ. **26**(5), 889–919 (2004). doi:10.1016/j.eneco.2004.04.016s
IDL: The future of geothermal energy. Idaho National Laboratory (2006)
IEA: World Energy Outlook 2010. OECD Publishing (2010). doi:10.1787/weo-2010-en
Jacobi, J., Starkweather, R.D.: Solar Photovoltaic Plant Operating and Maintenance Costs. ScottMadden Management Consultants, New York (2004)
Krohn, S., Morthorst, P.-E., Awerbuch, S.: The economics of wind energy, European Wind Energy Association (2009)
Ladanai, S., Vinterbaeck, J.: Global potential of sustainable biomass for energy. In: SLU, Institutionen for energi och teknik, Swedish University of Agricultural Sciences Department of Energy and Technology, Report 013 ISSN 1654–9406 (2009)
Lund, H., Hvelplund, F., Østergaard, P.A., Möller, B., Vad Mathiesen, B., Andersen, A.N., Morthorst, P.E., Karlsson, K., Meibom, P., Münster, M., Munksgaard, J., Karnøe, P., Wenzel, H., Lindboe, H.H.: Danish wind power export and cost (2010). ISBN 978-87-91830-40-2
McKendry, P.: Energy production from biomass (part 1): overview of biomass. Bioresour. Technol. **83**(1), 37–46 (2002). doi:10.1016/S0960-8524(01)00118-3
Mork, G., Barstow, S., Kabuth, A., Pontes, M.T.: Assessing the global wave energy potential. Shanghai, China, June 2010. In: Proceedings of OMAE 2010, 29th International Conference on Ocea
Nijmeijer, K., Metz, S.: Chapter 5 Salinity gradient energy, vol. 2. Elsevier (2010). doi:10.1016/S1871-2711(09)00205-0
OECD/Nuclear Energy Agency: Projected costs of generating electricity (2010) doi:10.1787/9789264084315-en
REN21: Renewables 2011 global status report. REN21 Secretariat (2011)
REN21: Renewables 2012 global status report. REN21 Secretariat (2012)
REN21: Renewables 2013 global status report. REN21 Secretariat (2013)
Royal Belgian Academy Council of Applied Science: Hydrogen as an energy carrier, April 2006
Schmidhuber, J.: Chapter ten—impact of an increased biomass use on agricultural markets, prices and food security: a longer-term perspective. In: CFE Conference Papers Series No.2. Center for European Studies at Lunds University (2008)
Sclater, J.G., Jaupart, C., Galson, D.: The heat flow through oceanic and continental crust and the heat loss of the earth. Rev. Geophys. **18**(1), 269–311 (1980). doi:10.1029/RG018i001p00269
Thorpe, T.W.: An overview of wave energy technologies: status, performance and costs. Wave power: moving towards commercial viability, Westminster, London, November 1999
United States Census Bureau: World population: 1950–2050, December 2008. http://www.census.gov/population/international/data/idb/worldpopgraph.php
US Department of Energy: Biomass for electricity generation. April 2011. retrieved from http://www.wbdg.org/resources/biomasselectric.php on 7 Feb 2012

Chapter 5
Outlook

Abstract The technological and economical aspects of the most important fossil-, nuclear- and renewable-based technologies have been reviewed in the preceding chapters. Presenting these technologies separately was necessary since they rely on different physical concepts. The drawback of such an approach is that it makes it challenging to compare these technologies against each other. The present chapter aims at correcting for this drawback by bringing all technologies together. A technique used to predict future energy costs will then be introduced and future electricity generating costs will be forecasted for onshore wind power and solar PV. Policies aiming at facilitating the deployment of renewable energy will be discussed before we conclude this book by looking into the technical and economical feasibility of supplying the world with renewable energy by 2050.

Keywords Energy Policy · Energy Spending and Efficiency · Price Forecasting · Future Global Energy Supply

5.1 Summary of the LCOE of All Technologies

Table 5.1 summarizes the costs of all the electricity generating technologies described in the previous chapters. A comparison solely based on the LCOE immediately indicates that the less mature electricity-generating technologies (c) such as solar, biomass, small hydropower and wind are typically more expensive than technologies relying on fossil-fuels (a), whereas alternative non-fossil based technologies (b), including geothermal, large hydropower and nuclear power are often cost-competitive with fossil-based technologies.

It has to be noted that the median cost is not fully representative of the difference in costs between the various technologies, especially for those which costs are primarily influenced by their location. At the same time, even the cheapest case of solar power project is not competitive with coal or natural gas for instance.

P. A. Narbel et al., *Energy Technologies and Economics*,
DOI: 10.1007/978-3-319-08225-7_5,

Table 5.1 Levelized costs of electricity for all technologies, ranked from cheapest to most expensive

Technology	Class	Median cost in Euro/MWh
Large hydropower	b	30
Near-surface geothermal	b	34
Black coal all	a	44
Black coal supercritical	a	45
Natural gas CCGT	a	53
Black coal with CC(S)	a	56
Natural gas all	a	55
Nuclear all	b	57
Solid biomass	c	61
Nuclear PWR	b	69
Wind onshore	c	86
Wind offshore	c	124
Small hydro	c	129
Solar CSP	c	165
Solar PV	c	232

Therefore, investigating the median price in this chapter is considered to be good enough. The main implication from Table 5.1 is that in the absence of some type of policy instrument supporting less mature technologies, these technologies would have difficulty diffusing into wider use in most cases. Amongst the 'non-competitive' technologies, onshore wind, biomass and some cases of small hydropower are closer to grid-parity on a pure cost basis than the existing solar and offshore wind technologies.

5.1.1 Experience Curve

Costs detailed in Fig. 5.1 may not be representative of future energy costs, because economies of scale, upsizing of technologies and learning effects (Edenhofer et al. 2011) via increasing volumes will lead to reduced costs over time. It was already in the 1930s that hints on a relationship between costs and cumulative production were identified (Wright 1936) and in the mid-60s, the Boston Consulting Group named this relationship: the *experience curve*. A technique relying on learning rates and average growth rates was then developed to draw forecasts on future costs of many products, processes and technologies (Ferioli et al. 2009). Such curves are decreasing linear relationship between the double-logarithmic scales of unit costs and cumulative volumes. The experience curve is used by many to justify support towards renewable energy generating technologies.

Computing the cumulative production of solar photovoltaic (PV) modules and plotting it against the cost per Watt produced gives Fig. 5.1. The experience curve is clearly visible.

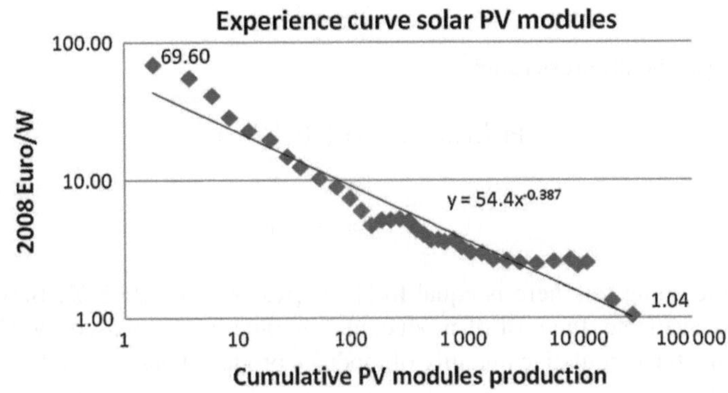

Fig. 5.1 Relationship between cost per watt produced and cumulative installed solar PV capacity (EPI 2007; Edenhofer et al. 2011)

The equation related to the experience curve is of the type:

$$Y_X = a \cdot X^b \tag{5.1}$$

where:

Y_X is the cost of the xth
X is the cumulative number of units produced
a is the cost required to produce a unit at a starting point
b is the slope of the function when plotted on a double logarithmic scale

and:

$$b = \log \cdot (\text{Progress ratio}) = \frac{\log(\text{Progress ratio})}{\log(2)} \tag{5.2}$$

The equation can directly be used to obtain the learning rate (1-Progress ratio), which describes by how much the production cost decreases every time the volume produced (or installed capacity) doubles.

Calculating the learning rate of solar PV module
Using Fig. 5.1, the relationship between the production and the cost of solar modules is given by equation $y = 54.4 \cdot x^{-0.387}$, where b, or the slope of the function when plotted on a double logarithmic scale is equal to -0.387. So, using Eq. (5.2):

$$b = -0.387 = \frac{\log(\text{Progress ratio})}{\log(2)} \tag{5.3}$$

we can get the progress ratio:

$$\begin{aligned}\text{Progress ratio} &= (10^{b\cdot\log(2)}) \\ &= 2^b \\ &= 76.5\%\end{aligned} \quad (5.4)$$

The learning rate here is equal to (1-progress ratio) = 23.5 %. Based on the data available, the cost of producing a module has decreased by 23.5 % everytime the cumulative quantity of modules produced has doubled.

Learning rate estimates for various energy generating technologies are reproduced in Table 5.2. The presence of substantial variation in the learning rates estimated for a single technology is a known issue (McDonald and Schrattenholzer 2001), which arises from the use of different datasets, geographical areas or dependent variable. This variation is a strong indication that learning rates must be treated with care. It however emerges from Table 5.2 that conventional energies have smaller learning rates than renewable energies, with the exception of large hydropower. A small learning rate is generally associated with mature technologies.

From a strictly objective perspective, the experience curve indicates that the cost of a technology is likely to decrease in the future, thus it justifies pursuing the support schemes in place. However, it cannot be used to predict with certainty when (or if) a technology will reach grid-parity,[1] since the predictions are based solely on history and do not include essential components such as forecasted technological improvements or planned scarcity of raw materials for instance.

5.1.2 Forecasting Future Energy Costs: The Cases of Solar PV and Onshore Wind Power

With the exception of large hydropower and near-surface geothermal, renewable energy technologies are relatively new in comparison to fossil-based technologies. Both active solar technologies and onshore wind power are no exceptions. New usually relates to limited experience for technologies that are not yet mature, which results in high prices. As producers, project developers and basically anyone involved in the solar or wind industry increase their experience with these technologies, redesign the systems, use more efficient and cheaper materials and as economies of scale occur due to larger volumes, the costs rapidly decrease. Costs given in Table 5.1 are thus a snapshot of the costs of these technologies today. They are much lower than what they used to be even five years ago, and they are likely to be higher than what they will be in the future.

[1] Grid parity means that an energy technology can compete without supports with the other energy technologies in a market.

Table 5.2 Learning rate estimates for various technologies

Learning rate estimates		
Technology	Period considered	Learning rate (%)
Biomass CHP	1990–2002	8–9[1]
Biomass		5[2]
Coal, oil and lignite		4[3]
Supercritical coal	1990–1998	4.8[4]
Combined cycle gas turbine	1990–1998	3.3[4]
Onshore wind power		15[2]
Onshore wind power		8[5]
Onshore wind power	1980–1998	15.7[4]
Offshore wind power		20[2]
Offshore wind power	1994–2001	8.3[4]
Solar PV		20[2]
Solar PV		21[5]
Solar CSP		10[2]
Solar CSP		7[5]
Nuclear		0[2]
Large hydropower	1980–2001	1.96[4]

[1]Edenhofer et al. (2011)
[2]Neij (2008)
[3]Kahouli-Brahmi (2008)
[4]Jamasb (2007)
[5]Timilsina et al. (2012)

Table 5.3 Average yearly growth of the cumulative installed capacity of solar PV and onshore wind power (EPIA 2012; WWEA 2012)

	Solar PV	Wind
2001	1,790	24,322
2002	2,261	31,181
2003	2,842	39,295
2004	3,961	47,693
2005	5,399	59,024
2006	8,980	74,122
2007	9,492	93,927
2008	15,855	120,903
2009	22,900	159,766
2010	39,700	196,653
2011	67,350	237,669
Av. growth rate	45.4 %	25.4 %

Learning rates can be obtained by using the methodology presented in Sect. 5.1.1. In addition, average annual growth rates are needed before forecasts can be drawn. The cumulative installed capacity for solar PV and onshore wind power for the period 2000–2011 is illustrated in Table 5.3.

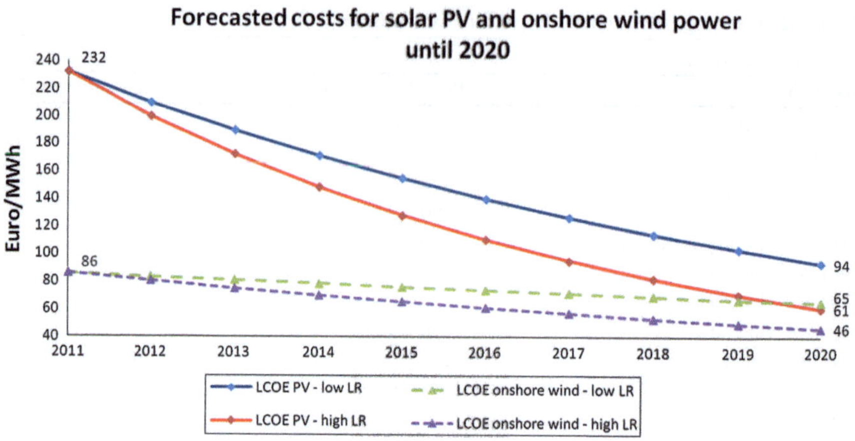

Fig. 5.2 Predicted levelized costs of electricity for solar PV and onshore wind power until 2020

The average growth rate over the whole period is the average of the annual growth rates. In the case of solar PV, a growth rate of 45.4 % indicates that the global cumulative installed solar PV capacity has increased by 45.4 % each year in average between 2001 and 2011. By combining the data from Tables 5.1, 5.2 and 5.3, it is possible to make predictions on the future costs of both solar PV and onshore wind power by using the following formula:

$$Y_{X_t} = LCOE_{2011} \cdot \left(\frac{X_t}{X_{2011}}\right)^{b_t} \tag{5.5}$$

where b is the slope of the relationship between cumulated capacity and cost, and X is the cumulative capacity at time t. The results are illustrated in Figs. 5.2 and 5.3.

Due to higher growth rates and learning rates, solar PV is predicted to experience a decrease in costs faster than onshore wind. However, solar PV starts from a much higher level and is not expected, based on this simple analysis, to achieve grid-parity with the cost of onshore wind power before 2020 in the best case. Three important simplifications/shortcomings of this analysis have to be underlined:

1. average growth rates are stable throughout time
2. all the components of a system are assumed to experience the same growth rate
3. the analysis is based on past information only

First, there is no economic reasoning that could defend the hypothesis that the average growth rate is constant throughout time for any technology. Taking a longer perspective, with a constant 25.4 % average growth in installed capacity, 3.5 TW of onshore wind capacity will have to be installed in 2030 alone. 3.5 TW is the equivalent of the capacity of nearly 160 Three-Gorges-Dam in China. Nonwithstanding the tremendous challenge it represents in the whole supply chain, there is no guarantee

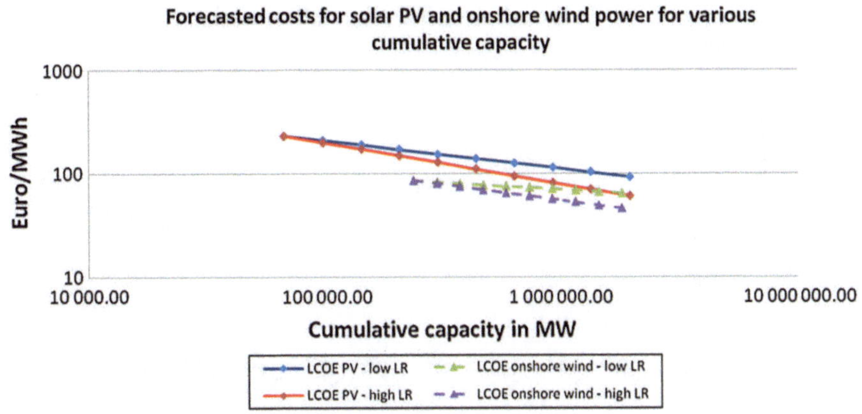

Fig. 5.3 Predicted levelized costs of electricity for solar PV and onshore wind power for various cumulative capacities (note the double logarithmic scale)

that economically suitable locations will be available for these wind turbines. A decreasing growth rate over time would thus be more appropriate than a constant growth rate.

Second, an energy system is composed out of several elements, all of which might have different degrees of maturity and will therefore experience different learning rates. For example, a PV system is composed out of a certain number of modules and the balance-of-system (BOS), which main component is the inverter. The latter serves to transform the direct current from the modules into alternative current that our electricity dependent devices are fond of. In terms of progress ratio, the correspondent median progress ratio for PV modules is estimated to be of approximately 20 % (EPIA/Greenpeace 2011; Edenhofer et al. 2011). Progress ratio for the BOS are approaching 20 %, whereas they are smaller, around 10 % for inverters (Edenhofer et al. 2011).

Third, as it was mentionned previously, solely including past events in making predictions might ignore serious concerns or expected improvments in the future. Figure 5.4 illustrates how investment costs have evolved over time in Denmark.[2] Until 2003, the investment cost is steadily decreasing, which tends to give support to the use of learning rates and growth rates to make predictions on future costs. However, no capacity was built in 2004. Worse, the investment cost is later increasing.

Leaving rough predictions for reality, this stepwise evolution of the investment costs in Denmark can be related to the evolution of the various support schemes in that country. Prior to 2004, each step can be connected to a change in support schemes (e.g. from subsidy to a premium on the market price), a decrease in support levels or to the suppression of a support scheme. Nearly no new onshore wind capacity was installed in 2004, which explains the gap in the time-serie. Some knowledge is necessary to understand the 2004 gap. Denmark has had very favorable policies

[2] Note that the experience curve cannot be derived from this figure since the volumes are missing.

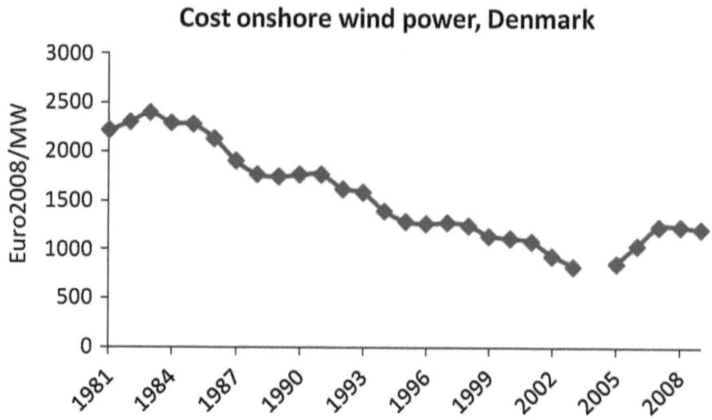

Fig. 5.4 Investment cost of onshore farms in Denmark Krohn et al. (2009)

towards wind power since the 1980s, which resulted in the installation of a vast number of wind turbines. So far, so good. However, Denmark has a naturally constrained territory which means that the optimal locations for wind generation purposes had all been taken by the early 2000s. Since these locations were taken first for obvious reasons, and since the first turbines were inefficient compared to newer models, best locations were utilized by inefficient models and the locations still available were not offering sufficient economical conditions for the installation of new capacity. Therefore, the Danish government initiated a new type of support that provided project owners with the possibility to scrap their wind turbines and receive a premium for doing so. Inefficient wind turbines were thus quickly replaced by newer models, which would not have been possible without this extra premium. The 2004 gap is therefore not the result of a decrease of the existing support scheme, but it is simply due to the planned suppression of a 'scrapping' premium. Without these combined schemes, no new project could be implemented economically that year. A new decommissioning premium was implemented in 2005 and the market started again. The period 2006–2008 was marked by a surge in raw material prices which made wind power temporarily costlier.

It can be concluded from this small story that even though it can be expected that the cost of wind power will further decrease in the future because of learning, the cost decrease will not be uniform around the globe and also that further cost decrease will depend on the availability to substitute currently used materials by alternative materials in times of scarcity. The expansion and renewal of favorable policy instruments until these technologies reach grid-parity will still be needed.

These shortcomings make the approach of using learning rates and growth rates problematic for long-term predictions but they typically play a minor role if the cost predictions are made only a few years ahead in time.[3]

---

[3] The existing literature Ferioli et al. (2009) claims that learning rates are appropriate for forecasting future prices correctly up to three order of magnitudes.

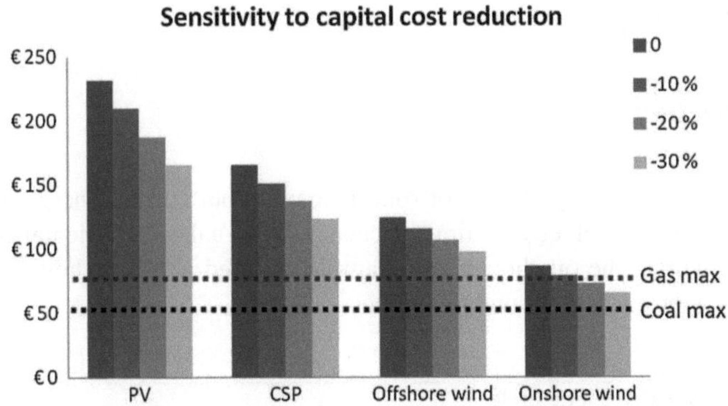

Fig. 5.5 Impact on the median LCOE of solar PV, solar CSP, offshore and onshore wind power of a 0–30 % reduction in capital costs (in Euro/MWh)

5.1.3 Sensitivity of Solar Power and Onshore Wind Power to Capital Cost Reduction

Based on the previous section, it is expected that the technologies with high learning and installation rates will achieve further cost reductions in the future. Nevertheless, it has not been shown what order of magnitude these cost reductions must be in order to make these technologies cost-competitive with conventional energy generating technologies.

This section precisely aims at illustrating this point and the case of solar PV, solar CSP and offshore- and onshore wind power are used as illustrations. For these technologies, it is clear that reduction in capital costs offer the highest potential since these technologies are capital-intensive. Figure 5.5 shows how the competitiveness of these four technologies is impacted with further reduction in capital-costs.

It can be concluded from the figure that even a 30 % reduction in capital costs will not suffice to bring most solar and offshore wind projects to grid parity with gas-fired power plants, let alone the case of coal. The story is different for wind and a 30 % reduction in capital cost will make onshore wind competitive with some gas-fired power plants and close to the most expensive coal power plants. A 30 % reduction in capital costs for onshore wind power shall take more time than for solar technologies due to lower growth rates, indicating that this technology is maturing.

Consequently, market intervention is needed in order to facilitate the deployement of renewable energy. Common market interventions are described in the following section.

5.2 Energy Policy

5.2.1 Introduction

Energy policies have played a major role in shaping our current energy landscape. These policies have evolved over time by cause of external events, mentality changes or other elements. The oil shocks, for example, marked a turn on how energy was perceived in many places around the globe.

Energy Policy in Denmark

In Dennmark, over 90 % of the energy consumed in 1972 was derived directly from oil. Domestic energy policies mostly promoted the exploration of domestic oil and gas reserves at that time (Krohn 2002).

With such a high dependency on oil, Denmark was badly hit by the 1973 oil shock and policies were reoriented towards energy diversification. Coal started to be heavily supported by the government. The second oil shock further impacted the country as, at that time, 78 % of all energy was still derived from oil and 20 % from coal. Coal was further developed in order to reduce the country's dependency on oil. As a result of these policies and higher energy demand, greenhouse gases emissions (GHG) in Denmark started to increase. Heightened awareness about possible climate change and local air pollution led to the establishment of an energy policy supporting the deployment of various renewable energies, mainly wind power. From a mere 2 % in 1979, the share of renewable energy in final energy consumption reached 23 % in 2011.

The case of Denmark illustrates the importance of energy policies. Such policies may pursue various goals and they all derive from specific problems. At present, GHG levels are increasing and it is assumed by many governments that these emissions enhance the possibility of experiencing a dramatic change in Earth's climate. Pursuing with the example of Denmark, recent governments have identified that energy generation, mostly based on fossil fuels, was/is responsible for a big share of Denmark's emissions. To allow the deployment of various renewable energy technologies, many countries, including Denmark, have implemented some type of specific policy instruments to facilitate the deployment of new renewable energy generating technologies.

Other potential goals pursued by energy policies include energy security (deriving from the dependency on politically unstable countries), energy diversification (derived from a dangerous dependency on a very limited number of energy sources) and industry creation (derived from a political will to create jobs and promote an industry). All the formentioned goals are the results of perceived market failures leading to situations that are not found socially optimal by governments, justifying governmental interventions in the energy market.

Once a government has decided to intervene in the energy market and promote certain energies, it still faces a vast array of implementable options. Policy instruments available to support the deployment of renewable energy can be classified in two categories: primary instruments and secondary instruments. Primary instruments include the tendering process, Feed-in Tariffs and Tradable Green Certificates (these are further detailed in Sect. 5.2.6), whereas secondary instruments include fiscal incentives, soft loans and other subsidies. If the latter category is also used in other sectors and for this reason, are generally well known, primary instruments are typical of renewable energy.

The need for market intervention in order to promote the deployment of new renewable energy generating technologies will first be introduced in this chapter. The concept of externality will be discussed next, followed by a description of the impact of a carbon tax on the cost of a technology. Prior to discussing the concepts of tendering process, Feed-in Tariffs and the Tradable Certificate scheme, a later section *Solving climate change via a carbon tax?* will review the likely causes of sub-optimal renewable energy policies leading to sub-optimal investments and poorly designed schemes.

5.2.2 Need for Market Intervention

It must be clear to the reader that less mature renewable energy technologies have not achieved grid-parity with conventional and more mature technologies. The aim of this section is to illustrate why market distortions (e.g. subsidy) are needed to support expensive forms of renewable electricity in a deregulated electricity market.

In a deregulated electricity market, buyers submit *bids* in advance stating how much electricity they want to purchase and at what price. At the same time, sellers submit *offers* stating how much they are willing to sell and at what price. Typically, sellers set their price based on their expected marginal costs in order to ensure that if their offer is retained, they will at least cover their operating costs (i.e. fuel costs and variable O&M costs). Offers and bids will then be ranked to create an economic merit order (supply and demand curve respectively). In Fig. 5.6, the market clearing price will be set at the intersection of the supply and the demand curve, which is denoted by p^0. This setting is representative of a market dominated by thermal plants (coal, gas, nuclear).

Now, let us assume that an investor is considering to invest in a wind farm and that a minimum average price over the lifetime of the project equal to $p^w = LCOE$ is needed in order to make this investment interesting. With a low marginal cost, the introduction of wind will change the economic merit order (see Fig. 5.7), shifting the supply curve to the right by the quantity of wind o^W. The clearing price (i.e. price at which the market is cleared) will shift down to p', short of what is needed to make the investment profitable.

In this setting and unless the wind producer receives some type of subsidy, wind power will not be built. Progress towards reaching grid-parity requires that these technologies get additional support, achieve significant cost reductions and/or that the cost of the other technologies goes up (e.g. increase in fuel prices or via a carbon tax for instance). Favorable policy instruments consequently have an important role to play in helping costly technologies reach grid-parity with conventional technologies, especially in the short term when technological innovation or increase in fuel prices is unlikely.

5.2.3 Externalities

Up to this stage, most of the emphasis was on the direct cost of energy and the concept of externality has been left aside. Externalities are by-products of the use of energy and are typically unwanted. '*An externality exists when the consumption or production choices of one person or firm enters the utility or production function of another entity without that entity's permission or compensation*' (Kolstad 2000). Applied to energy, the previous definition relies on two important elements: (a) the consumption or production of energy results in a cost to another entity and (b) this cost is not compensated for.

Based on the LCOE approach used to evaluate the direct cost of energy, it appears that electricity generation from coal is amongst the cheapest since only hydropower and near-surface geothermal can be more competitive in some specific settings. Nevertheless, important elements are missing from the picture, i.e. the externalities. When coal is burned to release heat, carbon dioxide, small particulate matter and mono-nitrogen oxides (NO and NO_2), among others, are discharged at the same time. Carbon dioxide contributes to increased concentration of that molecule in the atmosphere, a process believed to contribute to global warming. Small particulate matters have an impact on health and can lead to premature death. Finally, NO and NO_2 result in the formation of smog and acid rain. If not controlled, these by-products are externalities of the process of releasing heat from the combustion of coal. The cost of these externalities is therefore paid for by current generations

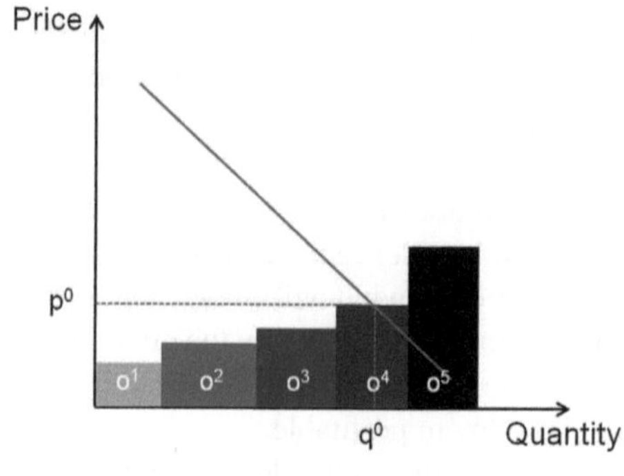

Fig. 5.6 Setting price in an electricity market

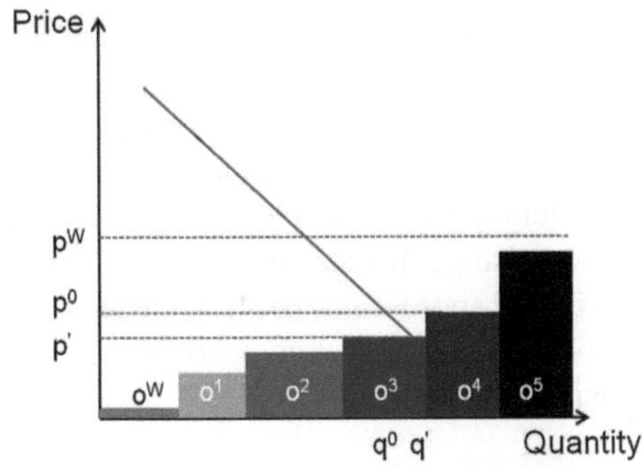

Fig. 5.7 Setting price in an electricity market with wind

via their health, decreased biodiversity and agricultural output and also possibly by future generations since they may have to cope with a warmer climate.

There exists a multitude of externalities. Wind power is often critized because some poorly located wind turbines cause the death of birds, noise and/or visual pollution. The externalities can also be positive. For example, offshore wind power can potentially create a sanctuary for fish since fishing boats are not allowed in offshore wind areas.

Quantifying precisely these costs is challenging and will often be subject to disagrements. However, it is possible to get an idea of the cost of 'internalizing' these externalities by estimating how much it costs to suppress them (Roth and Ambs 2004). For example, end-of-pipe measures can capture the small particulates before they are released into the air, the cost of which is known. Another approach is to 'internalize' these externalities by politically chosing what level of externality is acceptable (cap) and let a market regulate the price necessary to ensure that this cap is not crossed (trade). Such system is known as cap and trade system, of which the European emission trading scheme (EU-ETS) is a good example.

5.2.4 Sensitivity of Selected Technologies to a Cost on Carbon

Global on-going discussions under the United Nations Framework Convention on Climate Change (UNFCCC) aim at mitigating climate change by slowing down the increase in concentration levels of greenhouse gases in the atmosphere. A path to achieving this target is to create a market-based carbon tax in order to internalize the cost of emitting carbon dioxide. Under such scheme, governments agree on a maximum level of carbon that can be emitted by the industries falling under the system. The industries that emit carbon have to purchase the right to emit, which eventually make carbon-intensive technologies less competitive compared to carbon-free technologies, favoring the transition towards more efficient carbon-based technologies and/or non carbon-based technologies. The term *competitive* is perhaps missused here, because internalizing costs reflect the 'true' social cost in reality (if such concept exists).

Researchers (Helm 2012) tend to believe that a global carbon tax is the best way forward in tackling climate change, because a carbon tax does not pick winners, rather letting the market choose which technologies will minimize the cost of tackling climate change. It is deemed efficient as the cheapest options will be exploited first. The LCOE approach can be modified to evaluate the impact of a market-based carbon tax on the level of cost-competitiveness of the various technologies. An example is provided thereafter.

Application: How to calculate the impact of a carbon tax on the cost of electricity generated from a coal-fired power plant

A carbon tax is usually given in monetary units per tonne of carbon dioxide. The approach detailed hereafter for the case of a coal-fired power plant is an illustration of how a 20 Euro/tCO_2 carbon tax impacts the cost of that energy source.

The calorific value of coal is between 17 and 34 MJ/kg (see Chap. 1). Here we assume a carbon content of coal of 34 MJ/kg (which corresponds to anthracite). Now, carbon dioxide (CO_2) is composed of two atoms of oxygen and one atom of carbon. The atomic mass of carbon is about 12 unified atomic mass units (amu), whereas the atomic mass of an atom of oxygen is about 16 amu. Burning one kilogram of carbon thus results in $(12 + 2 \cdot 16)/12$ kg of CO_2. Consequently, burning 273 kg of coal will be accompanied by the release of 1 tCO_2. The heat energy of these 273 kg becomes:

$$E_{heat} = 34 \frac{\text{MJ}}{\text{kg}} \cdot 273 \text{ kg} \approx 9.3 \text{ GJ}. \tag{5.6}$$

The end user will receive a smaller amount of energy, which will be determined by the efficiency of the power plant. Again, from Chap. 2, we learnt that realistic efficiencies for coal based power plants are about 40 %. 273 kg of coal will therefore produce 1 tCO_2 and only 3.7 GJ of electricity.

Using an appropriate conversion factor between GJ and MWh (1 MWh is equal to 3.6 GJ), the carbon tax transforms as 20 Euro/tCO_2 $\longrightarrow$ 20 Euro/3.7 GJ = 19.5 Euro/MWh. Without the tax, the median LCOE for a coal based power plant amounts to 44 Euro/MWh. With the carbon tax (and assuming that a carbon content of 30 MJ/kg is representative of the median coal power plant), the LCOE increases to 63.5 Euro/MWh; or an increase of nearly 50 %.

This approach can be applied to different carbon taxes and different technologies. Figure 5.8 summarizes the results of this approach and clearly shows that the cost competitiveness of carbon-free technologies increases with increasing carbon taxes,

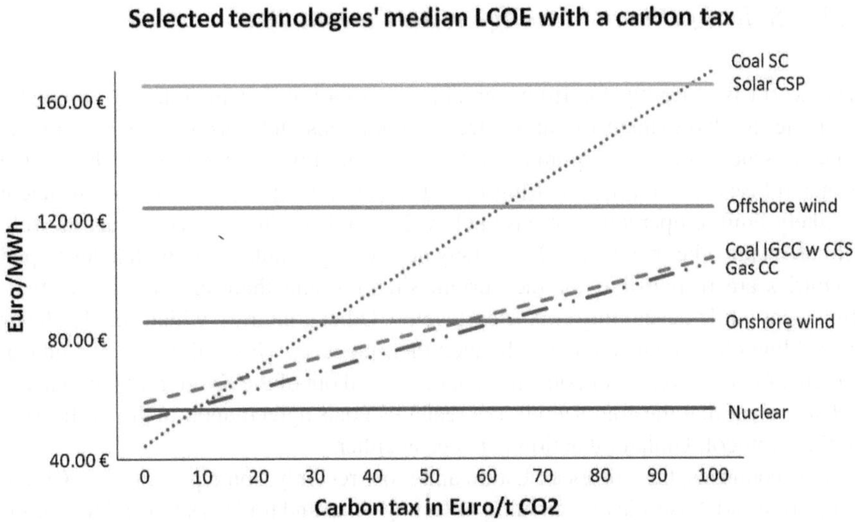

Fig. 5.8 Impact on the median LCOE of coal, coal with CCS and natural gas of a carbon tax ranging from 0 to 100 Euro/tCO_2. The median cost of solar PV is 231 Euro/MWh, it thus does not appear on the graph

since the cost of clean technologies will not directly be affected by the tax.[4] The cost of coal with CCS does not increase at the same rate as the cost of coal or even natural gas since the quantities of carbon dioxide released per unit of energy generated are lower in the case of CCS. The different slopes are due to different thermal efficiencies.

The European Union has implemented a system which puts a price on carbon for certain sectors: the EU-ETS (emission trading scheme). As of mid-2013, the cost of purchasing the right to emit one tonne of carbon dioxide was lower than 4 Euro/tCO2 in the European Union. At that level, natural gas is becoming increasingly competitive with coal, and some cases of wind reach grid parity with the most expensive coal power plants.[5] Such price-levels are however largely insufficient to promote solar technologies and even a price of 100 Euro/tCO2 would not suffice in their case. It must be duly noted that large variations take place between the LCOE of various solar PV projects. Therefore, ideally located solar PV projects may benefit rapidly and directly from the introduction of a carbon tax. However, solar PV projects located in countries with poorer solar conditions are significantly less likely to benefit from a carbon tax.

[4] We assume a carbon content for natural gas of 56.10 kgCO_2/GJ and that CCS leads to a 70 % reduction in the emission levels from coal, since a big share of the carbon dioxide is captured before it is released in the atmosphere.

[5] The reader will realize this if he/she makes the same analysis for the min/max prices of coal and wind.

5.2.5 Solving Climate Change via a Carbon Tax

The world is currently far from agreeing on a carbon tax in order to efficiently promote the deployment of carbon free technologies and thus address the climate change issuc. As a short summary of the status of climate negotiations, the United States refuses to commit to cutting its emissions in a way that would be sufficient to likely limit temperature increase below 2 °C unless China agrees to act as well. China, on the other hand, and along many developing countries claim that developed countries are responsible for the current situation and therefore should cut their emissions while developing countries develop. Other countries, which acted in favor of tackling climate change have changed their position. This is the case of Canada, which, after discovering unconventional oil, pulled out of the Kyoto protocol when it became clear that the country will not reach its goals agreed upon in the ratification of this protocol. Unilateral action is not easy either.

Few countries, the European Union and more recently, some provinces in China, California and Australia are pursuing or testing a cap and trade system and have chosen to act unilateraly. Even though the targets are often limited in terms of ambitions, these systems suffer from design problems. For example, there is a fear that inputing a tax on carbon unilaterally will render some industries uncompetitive, forcing them to relocate to other countries where the production technologies in place might have higher emission intensities compared to home technologies. Such a tax may also lead to the loss of jobs in favour of other countries. Moreover, certain industries (e.g.: coal industries in coal endowed countries) may be strong enough to lobby for policies that will suit them rather than benefit the country's citizens. Existing systems therefore focus on a limited number of industries, sometimes distributing carbon quotas for free. And in order to promote additional investment in renewable energy as a solution to climate change, energy insecurity and/or to create a renewable industry, many countries have started to introduce additional policy instruments supporting specific technologies, or in other words, picking winners. Overall, an excess amount of quotas and the introduction of additional policy instruments means that carbon prices struggle to reach a level sustainable for promoting investment in new technologies, even though the cost of these policies may be tremendous. Efforts to set a minimum carbon price typically fail.

The main policy instruments implemented in practice to support the deployment of renewable energy will be introduced next.

5.2.6 Common Policy Instruments Used to Support Renewable Energy

In practice, three types of primary policy instruments are used. These are the tendering process, the Feed-in Tariffs and Tradable Green Certificates.

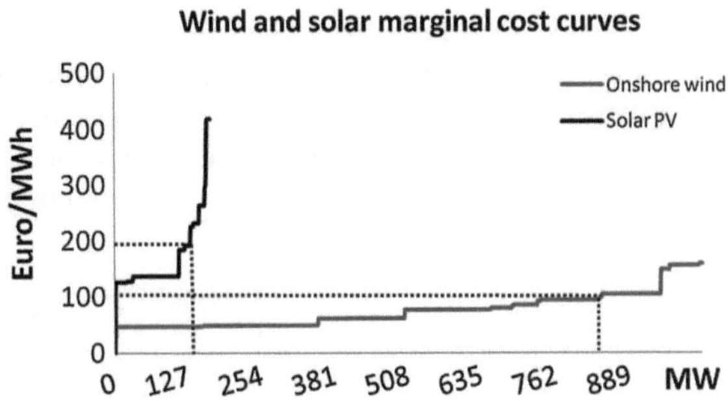

Fig. 5.9 Quantities obtained for different technologies, each with its own FiT

5.2.6.1 Tendering Process

A tendering process is a system in which a government will decide on a quantity of energy capacity to be built. Energy developers will have the possibility to bid a certain capacity at a price of their choice. The bids will then be ranked and the cheapest bids accepted by the government until the accepted bids sum up to the capacity wished for. In theory, this system is cost-efficient as the cheapest energy projects will be realized first. Yet, in practice, developers might be tempted of bidding a price that may not cover their costs with the hope of securing the project and the belief that they will access cheap capital and cheaper technologies in the future. There exists many examples of projects where such hopes have not been fulfilled and the projects abandonned. Example of effective tendering processes are the case of Denmark and the UK, and their successful reliance on a tendering process for offshore wind.

5.2.6.2 Feed-in Tariffs

Feed-in tariffs (FiT) are policy instruments guaranteeing a fixed price for each unit of electricity produced over a set period of time. In an electricity market, a FiT translates to a fixed price for each MWh generated and various technologies are often subject to different levels of FiTs to reflect their specific degree of maturity and costs.

A FiT is a price-based policy instrument in the sense that a government can choose at which level to set the tariff. Figure 5.9 pictures the case where a government would support two types of renewable energy (i.e. wind and solar) with two different levels of FiT. Under this hypothetical system, wind and solar capacity will be built until the marginal cost of installing more capacity equals the tariff. The deployment of solar panels and wind turbines is therefore completely independent of the regular energy price in the market.[6]

Investors therefore do not compete against the market price nor the other technologies. They simply need to ensure that the level of FiT applicable to their project

[6] Hence, investors do not have an incentive to build projects which deliver power when it is needed most.

suffice to cover their investment. Such policy instrument has the benefit of reducing uncertainty for investors because they know at which price their output will be sold at over a predefined period of time. For this reason, attractive FiT have resulted in the rapid installation of renewable capacity in numerous countries.

Under this strategy, the total profit experienced by the producer will be:

$$\pi^{p} = \sum_{i}^{n} ((P_i - C_i) \cdot Q_i) \tag{5.7}$$

where i refers to an energy technology, π^{p} is the total profit of the producers, P the price of the FiT, Q the quantity and C the cost experienced by the producer.

The total cost of this policy instrument S to the society becomes:

$$S = \sum_{i}^{n} (P_i \cdot Q_i) \tag{5.8}$$

This cost is traditionally paid via the use of cross-subsidies or by consumers through a tax paid related to the consumption of energy.

Prior to setting a FiT, policy makers need to estimate the marginal cost curve of a technology and unless they are very lucky (or unbelievably skilled), they will either overestimate or underestimate this marginal cost curve. In practice, the cost of a project is not known in advance with certainty (i.e. before a project is actually built). The uncertainty stems from the vague cost estimates of building capacity since many factors impacting the cost of a project are difficult to establish beforehand. For wind projects, many elements are not known with precision by policy makers choosing at what level to set the FiT. Local conditions (weather, average wind speed, topography or even oppositions from local residents) and the state and costs of a wind turbine at the time it is built by the project owner are such elements. Since the marginal cost curve cannot be estimated with precision, the total capacity built under a FiT and the cost of the scheme will only be known once the policy instrument is replaced or has expired. Figure 5.10 illustrates a case where the marginal cost curve is under- and overestimated by 50 %.

Cost of the Feed-in Tariff scheme

Let us assume that 1,600 MWh is generated from each MW of installed PV capacity on an annual basis. If the Feed-in Tariff is set to 200 Euro/MWh, the expected installed capacity will reach 130 MW (see Fig. 5.10) and the cost of the scheme for the first year is expected to amount to (using Eq. (5.8)):

$$130\,\text{MW} \cdot 200\frac{\text{Euro}}{\text{MWh}} \cdot 1{,}600\frac{\text{MWh}}{\text{MW}} = 41.6\,\text{MEuro} \tag{5.9}$$

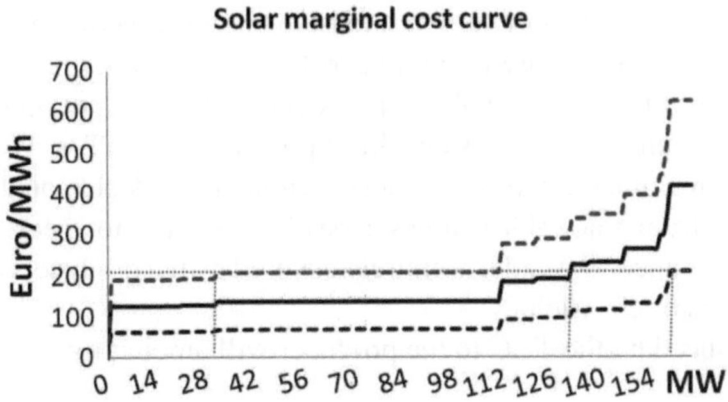

Fig. 5.10 Uncertainty in quantities of solar PV obtained under FiT if the marginal cost curve (+/− 50 %) cannot properly be estimated

Now, if the marginal cost curve has been underestimated by 50 %, only 30 MW of PV capacity will be installed. The cost becomes:

$$30\,\mathrm{MW}\cdot 200\frac{\mathrm{Euro}}{\mathrm{MWh}}\cdot 1{,}600\frac{\mathrm{MWh}}{\mathrm{MW}} = 9.6\,\mathrm{MEuro} \qquad (5.10)$$

Finally, if the marginal cost curve has been overestimated by 50 %, 165 MW of PV capacity will be installed. The cost becomes:

$$165\,\mathrm{MW}\cdot 200\frac{\mathrm{Euro}}{\mathrm{MWh}}\cdot 1{,}600\frac{\mathrm{MWh}}{\mathrm{MW}} = 52.8\,\mathrm{MEuro} \qquad (5.11)$$

It is now understable that FiT have resulted in a very high cost for countries that have underestimated the marginal cost curve M^C of an energy technology and in little capacity being built in countries that overestimated this M^C.

5.2.6.3 Tradable Green Certificates

Under a tradable green certificates (TGC) system, a producer will have the obligation to produce a share of its energy from renewable sources. For each MWh of green electricity produced, the producer will receive a set amount of certificates. Regularly, the producer will have to show that she complied with her obligation by providing these certificates to the competent authorities. If the producer lacks a number of certificates or if she has certificates in excess, she has the possibility to buy/sell these certificates on a secondary market, or to pay a fine. The rules of supply and demand will determine the price of the certificate and drive additional investment when needed. It is cost-efficient as investment will first occur where it is cheapest and leads to the replacement of carbon intensive energy generation by clean energy generation. Yet, the uncertainty in future certificate prices does not overly motivate investors. Because of this, renewable energy deployment does not occur as fast under a TGC system as under a FiT system. In opposition to a FiT, a TGC system is a quantity-based instrument. In this case, a government can decide on the quantity (or share) of energy from renewable sources to be obtained. The total cost of such policy instrument is here uncertain and it is entirely covered by the producers and eventually by the consumers of electricity.

The cost induced by the TGC to the producer will amount to:

$$C^p = \sum_i^n (C_i \cdot Q_i) \tag{5.12}$$

In the case the producer has the obligation to produce a share of her energy via renewable energy, her profit will be:

1. If production exceeds the producer's obligation, the producer can generate some profit by selling her green certificates in excess in the secondary market.

$$\pi^o = ((Q - O) \cdot (P^e + P^c)) - (C \cdot Q) \tag{5.13}$$

   where π^o is the profit made on renewable energy generated beyond the producer's obligation, Q the total production of renewable energy, O the quantity of renewable energy the producer has the obligation to produce, P^e the electricity price perceived by the producer, P^c the price of the certificate on the secondary market and C the production cost.
2. If the production of renewable energy equals the obligation, the producer will not be able to sell any certificates on the secondary nor will she have the obligation to buy any.
3. If production falls behind the producer's obligation, the producer will have to buy certificates on the secondary market to compensate for its unachieved obligation. Her total cost of purchasing these certificates will amount to:

$$C^o = ((O - Q) \cdot P^c) + C \tag{5.14}$$

If the supply of certificates exceeds the demand, the price of a certificate will decrease, thus slowing down investment in new capacity (exisiting capacity in a market suffice to fulfill the minimum requirements). If the demand for certificates significantly exceeds the supply, producers will realize that it would be less costly to invest in new capacity rather than purchase the certificates they need. The certificate's price thus serves as a trigger to accelerate/slow down investment in new capacity.

In practice, the producer will maximize profit and optimize how much conventional and green power should be generated given the tradable green certificate price. The interested reader will find a thorough discussions of the producer's objective function under TGC and FiT in Tamas et al. (2010).

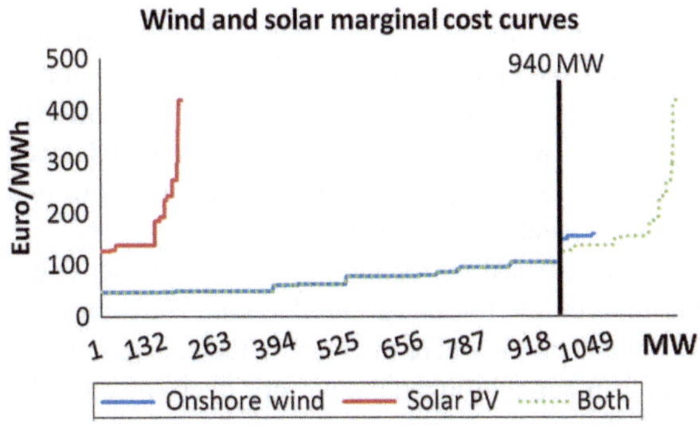

Fig. 5.11 Tradable green certificate schemes favor the instant cheapest option on a direct cost basis. In the case illustrated here, it means that unless over 940 MW of capacity needs to be built to reach the target, only wind farms will be constructed and benefit from the policy

Such system is *efficient* in the sense that the cheapest electricity-generating project will be built first as opposed to a FiT system where capacity of any technology will be built as long as it is profitable. The disadvantage of a TGC system is however that only the cheapest energy source is usually supported and less mature technologies do not benefit from such systems. In Fig. 5.11, unless over 940 MW of installed capacity is required, only wind power capacity will actually be built.

5.2.6.4 Choosing the Right Policy Instrument to Support Renewable Energy

Governments are facing a tough choice when deciding what policies should be implemented to support renewable energy. On one hand, a tradable green certificate scheme is cost-efficient whereas a FiT is not. On the other hand, an attractive FiT will lead to a rapid penetration of various energy sources whereas a TGC will traditionally favor only the most attractive energy source. A TGC may result in more GHG abatement than a FiT, but FiT will favor more investment in renewable energy. These are simple examples to show the difficulty in defining which policy instrument(s) should be implemented. The choice becomes even more complex if the targets defined by the government include the creation of a renewable energy industry, energy diversification or energy supply security. Also, countries' strategies are likely to be impacted by domestic resources as a country without access to oil or coal is more likely to be willing to become more independent than a country that has ample reserves for the next 100 years.

Focusing events such as the oil shocks, Tchernobyl or Fukushima also change the mindset of the population on various energy types which can impact the government's decision on which strategy to follow. Finally, even though FiT, TGC and tendering processes have been introduced, other policy instruments have been or are being used in practice, including the use of carbon taxes on specific sectors, investment subsidy, investment guarantee or RD&D grants.

In practice, there are as many supports schemes as there are countries. Switzerland introduced a FiT for a fixed amount of capacity. Denmark is supporting renewable energy with a combination of green taxes exemption and a premium perceived on top of the electricity price. Norway and Sweden have a common TGC system. Germany is using different FiT levels for wind energy projects depending on which area a project is based in and on how much capacity was installed the year before. The reason to have different FiT levels based on location is due to the fact that some locations are more suitable than others for energy production. Thus, the cost of generating power will differ depending on where the plant is located and different FiTs ensure that projects located in suitable locations are not over subsidized. These are few examples illustrating the variety of support schemes that have been implemented in practice.

5.3 Energy Spending and Efficiency

The purpose of the preceding chapters was to explain the possible ways of generating energy from various resources, the principles behind them, their limitations and their cost. The *necessary consumption*, or more precisely the transformation of energy from the given source to heat and other low quality forms of energy, will be considered in this chapter. Thus, we will answer the following two question: *How much energy do we need as a global society per unit of time?* and *What are the major sectors which contribute to this energy need?* When these questions have been answered, the alternative routes to reach the necessary global power production which are based on non-fossil alternatives will be considered.

5.3.1 Energy Spending

Table 1.4 illustrated the large variance in power per capita in various regions of the world, ranging from 9 kW (North America) to less than 2 kW (Africa). We shall assume that the North American power per capita is not a global natural goal. The average European consumption per capita is about 5 kW and we take for granted that the standard of living is fairly acceptable with a "5 kW European life". Consequently, we take this as a first estimate of a global level for every human being. But then, why not 4, 3 or as low as 2 kW per person on average? The latter is lower than the world average today, which involves large regions of poverty and high death rates. The low standard of living, including uncertainty and strain, experienced by humans at sub −2 kW conditions, generate instability and high birth rates. So this is clearly too low. In Europe, on the other hand, the population and energy consumption have been quite stable,[7] which is likely due to two facts: Relatively stable social

[7] At least from the 1960s and up to the financial crisis in 2008.

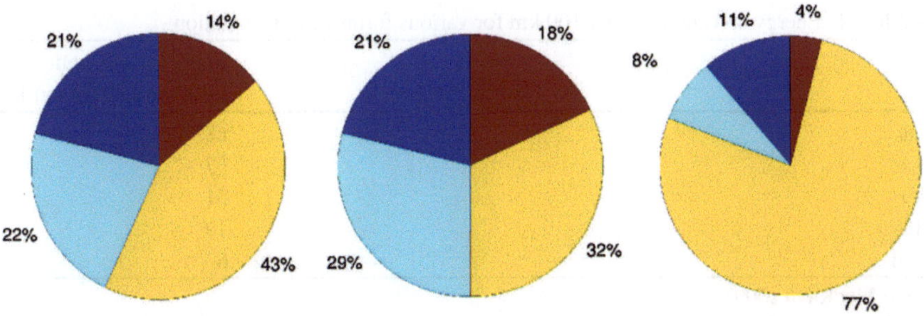

Fig. 5.12 Relative energy consumption by sectors in Europe (*left*), USA (*middle*) and China (*right*). The colors refer to transportation (*light blue*), residential (*dark blue*), commercial (*brown*) and industry (*yellow*)

conditions (distribution of energy) and enough energy to support the needs of the society. However, it seems likely that the European standard of living might be maintained at a lower power consumptions than 5 kW, still far above 2 kW. To make an estimate on how far below 5 kW it is possible to get, based on better technology and realistic energy savings, we need to consider in which sectors the major energy consumption takes place.

The major sectors of energy spending in US, Europe and China are shown in the pie charts in Fig. 5.12. The European and American spending charts appear relatively similar, which in fact, is characteristic of all western world societies. The four sectors in the pie charts can be merged into three energy spending groups, each taking about $1/3$ of the energy consumption: transport, industry and direct human spending—or "commercial and residential". The central question is then how much the consumption can be reduced within each group without significantly changing the 5 kW lifestyle. It should also be kept in mind that this question does not apply to regions of the world which are developing. Consider for example the final pie chart showing the distribution of energy spending in China. Here, the industry completely dominates since this sector is the driving force in the transition of China into a leading economic nation and behind the fast growth in average energy consumption per capita. At some point, this transition might be completed and the Chinese energy consumption per sector may resemble the one of Europe.

Each sector can be disaggregated as:

- Transportation
  This sector includes the energy spent on all forms of transportation, from car driving to long distance air or ship traveling.

  Transport in the public and industrial sector is subject to competition and it is likely that it is performed in a energy optimized way given the various needs of the countries. For example, the sushi in Japan is based on fresh fish, which implies that air transport has to be performed if the fish is produced a long way from the consumer. Consequently, reducing and changing the energy spending in the transport sector may have a large impact on our daily life.

Table 5.4 Energy consumption per 100 km for various forms of transportation

	Energy consumption kWh per 100 person-km
Car	68
Sea	57
Air	51
Bus	19
Rail	6

Data MacKay (2009)

However, changing the means of transport can lead to reduced energy consumption as shown in Table 5.4. This table indicates the energy needed per person per km traveled with various means of transport and it is clear that by traveling slightly less in general (implying that humans should live more in clusters), and using collective transport or bicycles instead of large private cars, the overall energy consumption will shrink. Further improvements can be achieved by using rail instead of airplanes for intermediate distance travels. If, in addition, people would use smaller cars instead of the typical 1,500 kg family car, the situation will improve even further. If a reduction of a factor of two seems overly optimistic, a reduction in the amount of energy spent on all forms of transportation in the order of 25 % seems possible. Such reduction can be achieved as a result of both new and more effective transportation technologies and a shift from less private to more public transport.

Energy consumption of buses and cars

The energy consumption of a car is a sum of the energy required for harvesting the raw materials, the production costs and the amount of energy a car spends on fuel during its life cycle. We could also subtract the energy gained from recycling of the car components when it is abandoned, but this part as well as the harvesting part will not be considered here. These two factors are the smallest in any case.

The production costs differ a lot between a combustion motor and a electric driven car. In a $\sim 10^3$ kg electric car about half of this at 2013 battery technology is batteries, i.e. 500 kg which is produced with an energy cost of 100 MJ/kg (Sullivan and Gaines 2010). The remaing parts of a battery car and the entire part of a combustion driven car is produced with an energy cost of about 30 MJ/kg. Both figures depend of course on the specific car model and battery type but as order of magnitude figures they should be ok. We then see that a 1,000 kg standard small car costs in the electric case about 65 GJ to produce while a combustion model costs only 30 GJ. So, from an energy point of view only, it is reasonable that a electric car should cost twice the price of a combustion engine car of the approximately same size.

The energy consumption of the two cars in fuel is of similar size but better for the electric car: The battery pack delivers around 30 kWh which allows you to drive ~100 km, so the energy consumption per km is 0.3 kWh/km. The energy consumption of a combustion car is about 2 MJ/km = 0.55 kWh/km. The origin of the almost factor of two difference is due to the efficiency of the electric motor as compared to the combustion engine (<40 %). Then we also know the range of the cars which is say 200,000 km. This gives a total fuel consumption of the combustion car of about 400 GJ which is more than 10 times the production energy. For the electric car it is slightly different about 200 GJ in spent fuel as compared to 65 GJ in production energy, i.e. about a factor of 4. But in both cases a clear majority in the energy consumption of a car is the fuel costs.

The European electricity price of about ~50 Euro^{2013}/MWh as compared to the typical 2013 gasoline price, ~1.5 Euro^{2013}/l = 1.5 Euro/40 MJ ~135 Euro/MWh imply that the fuel costs, given that you pay for it yourself, leads to a total fuel price of the electric car (if the batteries lasts!) of about 2,500 Euro while it is about 6 times more for the combustion car. The drawback of electric cars so far has been its limited range and the fuel time. But as soon as battery efficiency has doubled or tripled it should be the end of combustion car period!

A diesel bus has an energy consumption of about 5 liters per 10 km, which is about 10 times as much as the combustion car. From an energy point of view the energy per person km then requires 10 passengers in the bus on average to compete with the small private car, and if the private car owner replaces it with an electric car and always use it with 2 or more persons the bus cannot compete, as long as it is not electric.

- Industry

The industrial sector uses energy in a variety of ways. Often, energy is directly needed in order to raise the temperature of components or fluids, to induce chemical processes, etc. It is not easy to point out how this sector should use less energy without producing less units of whatever production we consider. The fact that industry pays a price for their energy input implies that more efficient production technologies will cut costs and thus be beneficial. Comparing the pie charts between China and Europe in Fig. 5.12, the industry sector is most likely to increase globally, because a much larger fraction of the people on Earth live in developing countries. Let us simply and realistically assume that higher energy prices and better production technologies may cut the relative spending in the industry sector by 10 %.

- Residential and Commercial

  Residential and commercial include each person's contribution from heating their homes and work, building roads, producing clothes, spending electricity, i.e. all things connected to human life.
  The largest contribution to this sector, ~ 60–80 % is the energy spent for heating and cooling purposes of living rooms, food, water, etc. In cold countries, energy is needed for heating during winter seasons while warm countries spend energy on cooling during summer. Reducing or increasing the temperature is often suggested as a way of saving energy. However, this measure is in conflict with the standard of living people want. For example, if an indoor temperature several degrees below 20 °C is suggested, not many Nordic citizens will find that acceptable. Thus, reducing energy spending in this sector is more likely to take place gradually, by replacing old houses with newer houses with better insulation and windows. Already today, so-called passive houses are built which are order of magnitude more efficient in energy conservation, storage and re-use of stored energy as compared to ordinary buildings. However, the construction of such houses is expensive and these do not manage to compete in the private housing market today.
  Now, increased loft insulation and new windows can reduce the energy consumption in existing houses by up to 30 %. We therefore assume that, in a 5 kW society of today, a decrease of 25 % is within reach in this sector, although the reduction will be obtained gradually as the result of a combination of a range of technology improvements and small changes in everday's life. Examples of the latter are slightly less consumption of heated water, slightly more vegetarian (and fish) as compared to meat and keeping our belongings slightly longer before we replace it.

Adding up the suggested saving potential of each sector leaves us with a realistic possibility to go from a total of 5 kW per person in Europe to about 4 kW per person.

5.3.2 *The Feasibility of Reducing Energy Consumption Through Efficiency Gain*

The feasibility of a transition from 5 to 4 kW not only depends on technology and consumption behaviors, but also on another important aspect or effect. The intuition behind this effect is first provided. Assume a home owner spends 100 monetary units on his heating bill every month. This home owner can decide to invest into improved insulation, in which case, the full heating cost will decrease by 30 %.[8] By choosing to invest into insulation, the owner saves 30 monetary units every month, at the same time as he reduces his energy consumption.

The story does not end here, because the owner is likely to spend these 30 monetary units on other goods. Let us assume that the owner can spend this saving either on

[8] "Full" is understood here as the levelized cost of heating a house.

watching TV or on driving a car. Each monetary unit spent either on TV, heating or in driving a car corresponds to a certain quantity of energy. Let us hypothetize that the energy content of a Euro spent on TV is lower than the energy content embedded in a Euro spent on heating, which, in turn, is lower than the energy content of a Euro spent on driving a car. In case the home owner spends his money on watching TV, his energy consumption will increase again, but by less than the initial saving of improving the house insulation. This is known as the *rebound effect*, i.e. energy savings in one category are offset by energy spendings in other categories. In the second case, where the home owner chooses to drive more, the increased energy spending will more than offset the energy savings of improving the house insulation. This is called a *backfire effect*.

The latter effect was first suggested by the British economist William Stanley Jevons in 1865, when he suggested that the more economical use of coal to do work increases the consumption of coal rather than saving it, which is known as the Jevons paradox. Rebound and backfire effects are particularly relevant as long as energy efficiency is thought of by governments to lead to lower consumption and lower negative environmental impact, as it is not necessarily true.

A third effect exists, which has not been named so far; the *negative rebound effect*. If a government forces the use of a technology which is more efficient and costlier than the technologies in use, it can be that there is a negative rebound effect. This is possible if the relative cost of using the technology, taking efficiency gains into account, increases. Consequently, efficiency gains must be compensated for in order to lead to effective energy savings, perhaps by introducing physical caps such as rationing or quotas. The decision of forcing emissions down has to be made carefully as certain industries may choose to relocate to foreign countries, which often have less energy efficient production technologies, in order to retain their competitiveness if the schemes put in place have poorly been designed. Assuming that consumers do not change their consumption bundle, the energy consumed per capita will effectively have gone up instead of decreasing.

Forcing emissions down in the oil sector in Norway

In an effort to reduce the country's emissions, Norway is pursuing a politic of electrifying its oil platforms. At the moment, most offshore oil platforms use they natural gas they produce to generate power, which they use to power their own operations. This process emits a large amount of carbon dioxide and it was thought that electrifying these platforms will allow the country to reduce its emissions significantly.

What this effort means practically is that subsea cable will be built between mainland Norway and the oil platforms, allowing electricity from hydropower to flow to the platform or alternatively offshore wind could be used to generate the electricity. The natural gas saved can then be exported to Europe. As a result, production-based emissions in Norway decrease. Yet, since the natural gas is burned anyway altough not in Norway, electrifying offshore oil platforms do not change the amount of emissions at a global level, unless this natural gas replaces coal generation.

5.3.3 Future Energy Alternatives

Here, we assume that a standard of living of 4 kW is a fair and feasible goal for all societies on Earth. With today's population of 7 billion people, a 4 kW goal implies a total global power of 28 TW as compared to the 2012 figure of about 16 TW when all primary bioenergy consumption is also included. If the population increases to 9 billion by 2030, the need for power automatically increases to 36 TW. When considering the changes from 1990 to 2008 (see Table 1.4), this is not a very likely scenario on a short timescale of 10–20 years, but it may be possible in 50–100 years. An example which illustrates the need for slow change is the following: For many families investing in a house is the main investment during the family "lifetime". Additional investments towards a passive house may double the capital costs. Thus increasing the energy efficiency of buildings is a process which needs to take several years or even an entire generation. Regarding private cars, it is less obvious: A new car only costs about 10 % of the total energy spending of buying and driving a car for 10 years. If you change your car very often, say every second year, you clearly spend more energy. But on the other hand you "support" the process of making more energy efficient cars, provided you do not buy an increasingly bigger every time you change.

The key question here is: *Based on which energy sources and which energy carriers can a global 4 kW per person society be achieved*? In some cases, the physical limitation on certain energy sources are indisputable (for example hydroelectricity), in other cases it is a limitation due to costs (solar power). Keeping known limitations in mind, a doubling of the global power consumption to 30 TW may be achieved by one or a combination of the alternatives presented below. Predictions on the cost of these options will be provided in the last section.

- *Fossil fuels with an increasing use of coal* This is today the cheapest and most likely alternative. The drawbacks are faster depletion of fossil fuel resources and increased greenhouse gas emissions.
- *Large scale buildup of nuclear energy* As one new nuclear energy plant can provide typically 3–5 GW of power the global need can be provided by 5,000–10,000

nuclear power plants. This number is 10–20 times larger than the number of existing plants. Four factors makes this alternative less probable than the former. (i) the present public opinion against nuclear energy, (ii) massive capital costs, (iii) the long term supply of ^{235}U and (iv) the much easier access to nuclear weapon material which would become available as a consequence of nuclear energy plants available "everywhere".

- *Large scale upbuilding of renewable bio, wind and solar power* This alternative requires massive investments as well. In addition, large land areas will have to be reserved. Since not all countries have enough space for their own renewable supply, special measures to secure energy supply would also need to be implemented. As any prediction regarding the future has an inbuilt uncertainty, so are these.

New technologies may open completely new alternatives. Some of these are;

- *New Nuclear* As discussed in Chap. 3, there are two alternatives. New fission reactor types which effectively can use ^{238}U or Th as fuel. The fuel problem would then be solved for thousands of years. In addition, the problems with nuclear waste and proliferation will be greatly reduced. The other alternative, nuclear fusion, would remove the waste and proliferation problem completely. Whether this is a real alternative remains to be answered in the current ongoing fusion research projects which aims to demonstrate that the technology works.
- *Unconventional fossils* As discussed in Chap. 2, the amount of available unconventional fossil fuel sources may well take over and prolong the lifetime of the fossil based timezone for another 100 years. Several alternatives are foreseeable, perhaps in combination. Shale oil and gas is currently undergoing explosive growth and heavy oil and tar sand projects are in developement. Finally, the exploitation of aquifers may also turn out beneficial.
- *Unconventional renewables* A last possibility is a breakthrough in our ability to take advantage of geothermal energy on a grand scale. Less likely to give a contribution are energy sources based on tides, waves and osmosis, but even these cannot be excluded.

As any prediction of future events, it is very uncertain which option is most likely. Key factors are present costs and technological limitations and development of the climate. In addition comes social factors like human demography and whether the development takes place without wars or other catastrophes.

5.3.4 A Renewable Alternative for Europe?

Let us now consider Europe and discuss which measures would be needed to obtain a fully renewable energy mix. When dividing the total area by the number of people, we find that each european citizen "owns" about 9,000 m^2. Due to low efficiency, we will neglect bioenergy as a significant energy source. We also leave out geothermal, which so far only works on a large scale in specific regions. This leaves us with hydroelectricity, wave/tidal power, wind power and solar power as the only contributing energy sources. Their various contributions to reach 4 kW per person can be estimated by the following;

1. Hydroelectricity: Assume that today's production may be doubled by building new power stations and upgrading the existing stations. This would then give an annual energy production of around 500 TWh or maximum 0.1 kW per person.
2. Wave and tidal power: Assume the entire Atlantic coastline of about 5,000 km is used for a gigantic wave power system which deliver 5 kW/m. This will in total give 25 GW or up to 0.05 kW per person if tidal power is also taken advantage of to some extent.
3. Wind power: Assume that 4–5 % of the European land area of about 10 Mkm^2 is filled with wind turbines delivering on average 2 W/m^2. This will give a total wind power production of about 1 TW. Assume that this can be doubled by a corresponding build-up of offshore wind farms. Then the contribution per person amounts to about 2 kW per person, a signicant fraction of what is needed.
4. Solar power: Assume that also 4–5 % of the European area can be transformed into massive solar power plants. 2 kW per person is obtained by transforming about 2 % of the European area into solar power plants. This 150,000–200,000 km^2 or 30–40 % of the area of Spain.

These numbers show that in a no-nuclear, no-fossil energy scenario for Europe, wind and sun may play the main role. Wave and tidal seems insignificant and hydropower can play only a small role. Turning this scenario into reality will imply that a total area up to the size of Spain is transformed into a power plant. This is possible, but not very realistic as long as fossil or nuclear alternatives exists with orders of magnitude higher energy concentration and lower price.

Several examples of countries which have already started to phase-out their fossil-fuels and of countries which have a great potential for renewable energy generation are provided next. Pointers on the reasons/limits which allowed/prevented this country to act will be indicated.

Practical example: Sweden

Figure 5.13 illustrates the evolution of the contribution of various energy sources to Sweden's total final energy consumption. It is noticeable that even though the Swedish population and gross domestic product (GDP) increased significantly during the time period considered, the level of final energy consumption remained virtually unchanged.

Sweden was largely relying on oil products for its energy needs before the oil crises. After the 1973 oil shock, Sweden started its nuclear program and the reliance on imported oil started to slowly decrease. The second oil shock at the end of the 70's triggered a biofuel era for Sweden and to large cuts in oil imports. Since then, Sweden has increased its reliance on clean

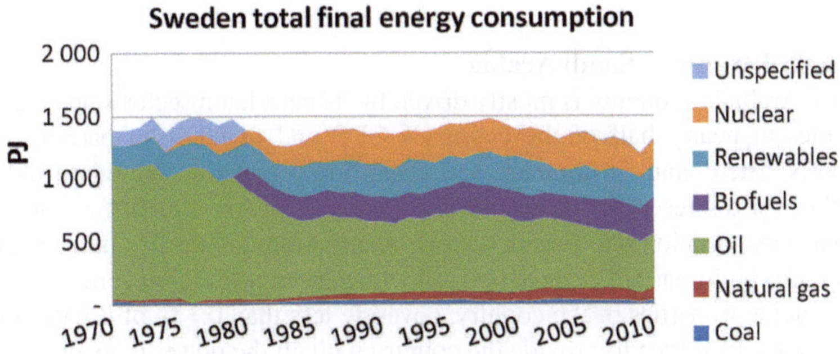

Fig. 5.13 Sweden's total final energy consumption (IEA 2012a)

energies generated domestically to the detriment of imported fossil-fuels and in 2012, the share of fossil-fuel was lower than half of the total final energy consumption.

Practical example: South Africa
At present, South Africa's power sector is largely dominated by coal with over 95 % of the power generated coming from coal. At the same time, South Africa is one of the sunniest place on Earth and the country's long coast line provides it with access to large untapped wind resources. Combining wind and solar (especially CSP) could allow South Africa to phase out its coal-fired power plants and create a decentralized energy generation system, thus solving its recurrent power outage issue and reduce the country's impact on the climate. However, a couple of factors may prevent such development. First of all, South Africa is a developing country and it can reasonably be argued that money should be used to empower the non-white (to increase equality after the Apartheid) and bring people out of powerty rather than on costly energy generating technologies. Second, South Africa has access to large and cheap coal reserves, therefore the gap in costs between fossil-based technologies and renewable technologies is higher in South Africa than in other countries, making the process even more difficult. Finally, large reserves of shale gas have been found, which is a cheap alternative to coal in case coal reserves were to be depleted.

Nuclear power could be an alternative for South Africa, especially because its reserves of Uranium would suffice to cover the country's energy needs for the decades and perhaps centuries, to come.

Practical example: Saudi Arabia
Saudi Arabia's economy is mostly driven by its petroleum sector since it contributes to nearly half of the country's GDP and to 90 % of earnings on exports. The country's immense and cheap fossil fuel reserves (about 20 % of the world's reserves) mean that the energy is nearly 100 % fossil-based (both transportation and power sector). Saudi Arabia is mostly desertic and the solar conditions are close to optimal with generally low cloud cover. Given the unique properties of the country, covering less than 0.1 % of it with solar panels would suffice to provide the country with all the power it needs.

Like the case of South Africa, the reader could imagine that Saudi Arabia would not want to go into renewables because of its large fossil fuel reserves. Nevertheless, Saudi Arabia has plans for renewable energy to supply for up to 10 % of its power by 2020. The reason why Saudi Arabia is looking into solar power for its power sector is straigthforward: oil is an inefficient way of generating power.

Let us look at how much it costs to generate power.

Current oil prices are around 100 USD/bbl, which corresponds to 65 ^{2008}Euro/bbl (i.e. to make it comparable with the other numbers in this book). Knowing that there is 0.146 tonne of oil equivalent per barrel, that there are 11.63 MWh per tonne of oil equivalent and assuming an efficiency of around 35 %, the cost per MWh from oil amounts to about 120–130 Euro/MWh, which is almost as expensive as the cost of offshore wind power.

It is obvious that the cost of the oil used by Saudi Arabia is much lower than the price on international market (a couple of USD/bbl only?). Nevertheless, there is a big opportunity cost involved in using the oil domestically rather than selling it to oil starving nations and this is where renewable energy plays a role. If generating power from the sun can compete with the cost of generating power from oil, it is likely that Saudi Arabia will phase out oil from its power market and export the precious fuel.

5.4 Estimating the Cost of Future Global Energy Supply

5.4.1 Introduction

In the final section of this book, we create three scenarios to evaluate the physical and economical feasibility of replacing fossil fuels with renewable energy and nuclear energy by 2050. This section starts by a discussion on the uncertainty associated with this type of work. The scenarios will then be further detailed, followed by a

Table 5.5 Comparison between forecasted and realized values for 2010.

	Forecasted (IEA 1994)	Effective (IEA 2012c)	Relative difference
World primary energy demand	11 560 Gtoe[1]	12 730 Gtoe	10 %
Coal (%)	24.5	27.3	11
Oil (%)	38.1	32.3	15
Gas (%)	23.2	21.5	8
Other (%)	14.0	18.9	35
Mean oil price	26 Euro^{2008}/bbl	115 Euro^{2008}/bbl	342 %

[1]is in million tonnes of oil equivalent

description of the approach used to forecast future energy costs. Finally, results will be presented.

This section is relevant to make the reader understand the use and limitations of the methods introduced in this book, as these are used to predict what the future energy supply might look like and how costly it might be to get rid of fossil fuels.

5.4.2 The Value of Developing Scenarios

Scenarios are highly uncertain as they depend on a vast amount of numbers. To show how exact forecasts can be, it is sometimes useful to look back at previous scenarios and highlight some of the elements that proved to be correct and those that were not. We do this here by comparing some of the values forecasted in the reference scenario of the International Energy Agency in its 1994 World Energy Outlook for 2010 (IEA 1994) and 2010 values.

Column *Relative difference* in Table 5.5 shows how erroneous forecasts can be (see for example the relative difference between the forecasted oil price and the effective oil price in 2010). Nevertheless, scenarios are useful in getting a general idea (after all, shares of primary supply provided by the various fuels were not "ugly") of what the future might look like. In order to be useful, these scenarios need to be updated whenever new information is available.

The scenarios we create in this chapter fall in that scope. They are merely here to provide a hint on whether we can physcially and economically replace fossil fuels with renewable energy and nuclear energy given our understanding of energy, the tools that are available to us and the uncertainty in predicting some of the variables needed to predict the future.

5.4.3 Scenarios

Three exploratives scenarios covering the period 2009–2050 are created to explore various possible futures: the *business as usual* scenario, the *renewable energy* scenario and the *nuclear* scenario.

The *business as usual* scenario predicts a future energy-mix incorporating major countries' current energy strategy and their recent past actions. Coal remains a key energy source throughout this scenario because it is among the cheapest source of energy in numerous countries and economically recoverable coal reserves will suffice to meet the demand for the upcoming decades. Unless externalities are included in the cost of energy, there is little reason to believe that the attractiveness of coal at a global level will decrease in the near future. Regarding the other fossil fuels, oil is expected to remain an important fuel, although the use of oil eventually declines due to limited proven reserves. The use of natural gas is forecasted to increase with the emergence of unconventional natural gas and potentially decreasing natural gas prices. Overall, the share of the TPES[9] from the various fossil fuels is expected to decrease from 81 % in 2009 to 70 % in 2050.

The second scenario is the *renewable energy* scenario and the aim in this future is to supply as much of the 2050 primary energy supply with renewable energy as possible. How much energy can be obtained from the various renewables depend on their potential. At the end of the renewable energy scenario, fossil fuels account for only 3 % of the TPES, due to the power locked-in (more on this later). The share of non-solar renewable technologies reaches 44 % and solar technologies are needed to match energy supply and demand, because of the other renewables' limited potential. By 2050, the total solar PV installed capacity reaches slightly less than 44 TW. Compared to the 2009 installed PV capacity of 21 GW, 44 TW means that the existing capacity at the end of 2009 needs to be multiplied by 2,000 times over a period of 40 years. Said otherwise, an average yearly growth rate of 21 % needs to be sustained over the following four decades, which remains significantly below the yearly growth rate of 2010, 2011 and 2012. Wind power (both onshore and offshore) is the second largest contributor to the TPES in 2050. Yearly installed wind capacity increases progressively to reach 480 GW of added onshore wind power capacity in 2050, which appears to be feasible because it represents 'only' ten times the new installed capacity in 2012. Biomass eventually contributes to 18 % of the TPES. Finally and even though their feasible potential is fully exploited by 2050, geothermal (4 % of the TPES) and hydropower (4 %) play a lesser important role.

A third option can be envisaged to address the issues of climate change and the rarefaction of fossil-fuel resources: nuclear energy. In the nuclear scenario, countries avoid using fossil-fuels by relying on nuclear power and renewable technologies. In this scenario, nuclear power supplies about 22 % of the total primary energy in 2050. The use of nuclear power limits the need for costly renewable technologies.

---

[9] TPES refers to the total primary energy supply. It is the amount of primary energy required to generate the energy used by the human society (the so-called FINAL energy supply).

Population, future GDP and TPES forecasts are common to all scenarios. Global population is expected to reach over nine billion humans in 2050 (United States Census Bureau 2008; UN 2003). GDP forecasts are obtained by using 2009 numbers (IEA 2011a) and assuming a steady 3 % growth rate. Estimates on the future TPES are obtained by combining population forecasts and forecasted growth rate in average power consumption per capita. Global energy consumption initially increases by 0.9 % and declines by one percent annually until the end of the scenarios in order to reflect a slowdown in the growth of the average power consumed per capita in the long run. Consequently, the TPES is expected to increase by 70 % over the time period considered and the average power consumption per capita increases from 2.4[10] to 2.8 kW between 2009 and 2050. Cumulative capacity is needed in order to apply the experience curve approach. It is assumed that the installed capacity at the end of 2009 (REN21 2010) is a good proxy for the cumulative installed capacity of the various technologies considered at the beginning of the analysis, even though it is known that some wind turbines and solar panels have already been decommissioned for instance.

Locked-in power exists in each scenario because existing power plants (and plants under construction) are kept in operation until the end of their economic plant life. At similar overnight costs, the cost per unit of electricity generated by a coal power plant will be several folds higher with a plant life of one year compared to a plant life of 40 years. It is therefore deemed not socially optimal[11] to replace the existing plants before the end of their economic plant life. The power generated from these power plants is considered to be *locked-in* until the end of the plants' economic life. The amount of locked-in power is slightly exaggerated to allow the renewable energy technologies to fill the gap between the locked-in power and TPES needed without being subject to unrealistic diffusion rates. The quantity of locked-in power is the same in all three scenarios and its costs is not part of the total cost of each scenario. The aim of this study is therefore to estimate the cost of the power needed to meet the gap between the quantity of power locked-in and the demand for primary energy in three different scenarios.

We use the median cost of the various technologies presented in the previous chapters as a starting point for our analyis and we assume a learning rate of 8 % for small hydro, of 10 % for solar CSP, of 15 % for wind energy and of 20 % for solar PV.

---

[10] In 2009, the TPES was equal to 16.3 TW (IEA 2011a) which is comparable to an average power consumption per capita of 2.4 kW.

[11] True unless externalities are included.

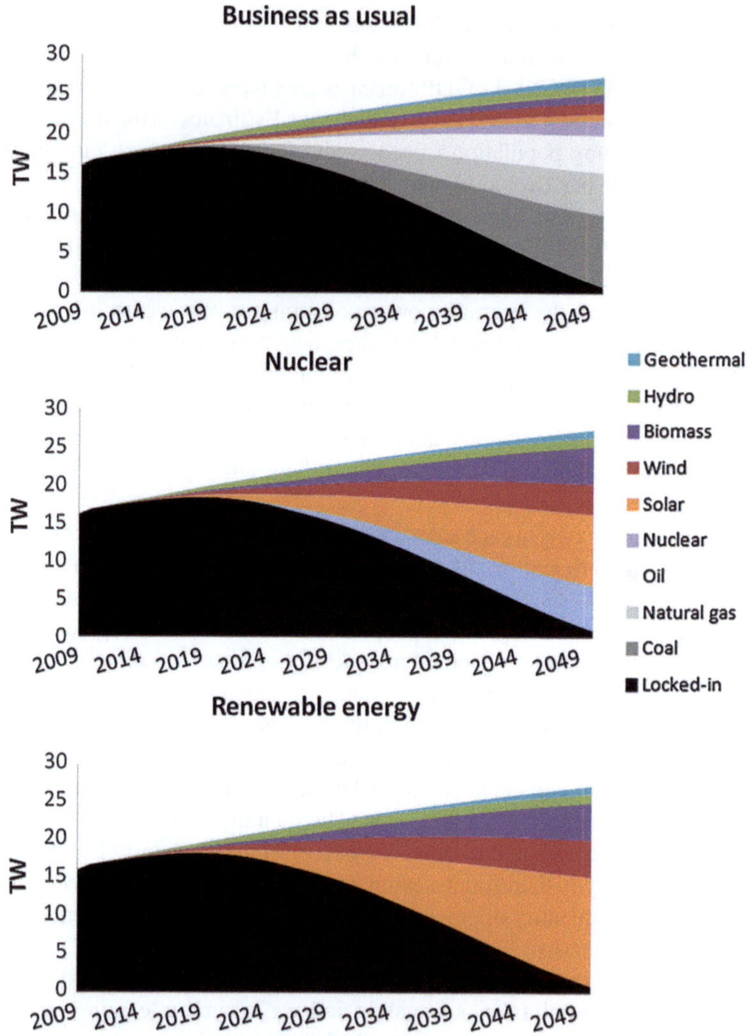

Fig. 5.14 Deployment of energy in the three scenarios

5.4.4 Forecasting Future Energy Costs

We model future energy costs using an inovative experience curve, which addresses, to some extent, the issue of ever-decreasing costs.[12] The equation used to model future costs in this study is detailed below:

[12] The use of experience curves is sometimes critizised as it assumed constant learning rates and building more thus always leads to lower costs.

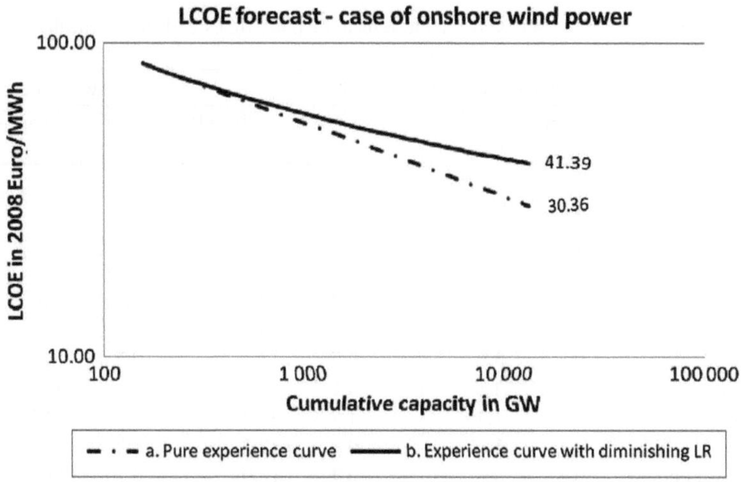

Fig. 5.15 Forecasted LCOE with: **a** Pure experience *curve* and **b** Experience *curve* with a diminishing learning rate

$$
\begin{aligned}
Y_t &= Y_{t-1} \cdot \left(\frac{X_t}{X_{t-1}} \right)^{b_t} \\
b_t &= \log_2(1 - l_t) \\
l_t &= l_{t-1} \cdot \left((1 - d)^{\log_2\left(\frac{X_t}{X_{t-1}}\right)} \right)
\end{aligned}
\tag{5.15}
$$

where l_t is the learning rate at time t and d is a rate at which learning diminishes. With this approach, the learning rate diminishes overtime to reflect that learning is more challenging to obtain in the long term. This rate is initially set to 10 % and a sensitivity analysis will later be provided to show the impact of changing this rate on the final cost of each scenario. The forecasted levelized costs of onshore wind power using this alternative approach are reproduced in Fig. 5.15.

5.4.5 Results and Discussion

Table 5.6 summarizes the costs of each scenario. The total cost for each scenario is the share of the cumulated GDP needed to fill the gap between the locked-in power and the total primary energy supply overtime.

Pursuing the *renewable energy* scenario would cost an extra 0.6 % of the cumulated global GDP between 2009 and 2050 compared to the cost of the *business as usual* scenario, whereas the cost of the *nuclear* scenario reduces this gap in cost by half. Part of this difference in costs can be traced to the rather extensive use of solar PV

Table 5.6 Cost of the three scenario for the selected technologies and their contribution to the TPES in 2050 (in %)

Technology	Business as usual		Nuclear		Renewable energy	
	Cost (%)	2050 TPES (%)	Cost (%)	2050 TPES (%)	Cost (%)	2050 TPES (%)
Coal	1.3	33.1	–	–	–	–
Crude oil	1.0	17.1	–	–	-	–
Natural gas	1.1	19.7	–	–	–	–
Nuclear	0.3	7.1	0.8	21.8	–	–
Large hydro	0.2	3.0	0.2	3.0	0.2	3.0
Small hydro	0.2	1.0	0.3	1.4	0.3	1.4
Biomass	0.2	3.9	0.7	18.1	0.7	18.1
Geothermal	0.1	3.6	0.1	3.6	0.1	3.6
Solar PV	0.2	1.5	1.7	21.9	2.2	32.5
Solar CSP	0.1	1.2	0.8	12.2	1.2	20
Onshore wind power	0.4	5.0	0.7	11.0	0.7	11
Offshore wind power	<0.1	0.7	0.2	3.8	0.3	7.3
Total	5.09	96.78	5.35	96.78	5.68	96.78

in the *renewable energy* and the *nuclear* scenarios, with the LCOE of solar PV not expected to go below 60 Euro/MWh in any scenario.

The difference in total cost between the scenarios may appear limited and this seemingly low cost of going fully renewable is mainly the result of two factors. First, the business as usual scenario also relies on renewable energy generating technologies to some extent (13 % of the TPES in 2050). From the approach chosen to forecast future LCOE costs, the first units installed are the most expensive. Because less renewable capacity is installed in the *business as usual* scenario than in the other two, it is the scenario where renewable energy will be the most expensive per MWh produced. The second factor is that fossil-based technologies are not subject to learning, whereas the other technologies are. By the end of the scenarios, it means that some renewable technologies get cheaper than some of the fossil-based technologies. For instance, wind power and biomass are expected to be competitive with natural gas by the end of all three scenarios. It is clear that there is more uncertainty in the long term than in the short term. For instance, the occurrence or not of the oil peak during the period considered brings uncertainty in the future fossil fuel costs. The real future costs of renewable energy generating technologies in forty years can also be debated at length. These two elements will be discussed further with the help of sensitivity analyses in the next section.

5.4.5.1 Sensitivity Analysis

Pressures on fossil-fuel prices will be stronger in the *business-as-usual* scenario than in the other two. Yet, fuel prices were assumed to be escalating at the same rate in

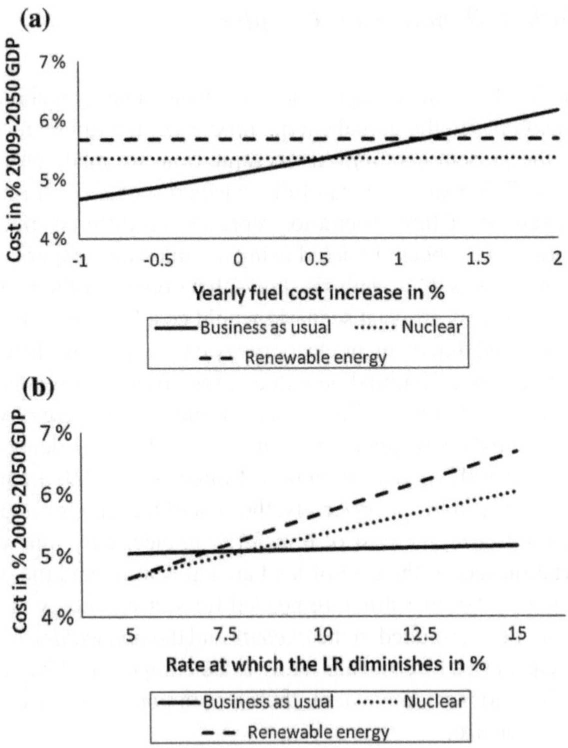

Fig. 5.16 Sensitivity analysis for different escalation rates of the fuel prices

all scenarios, because future fuel prices are very uncertain. A higher escalation rate, due to diminishing fossil reserves for example, reduces the gap in costs between the three scenarios and can even make the *business as usual* scenario less attractive than the other scenarios (see Fig. 5.16a). The opposite case where fossil fuel prices would decrease is possible too, provided that unconventional fossil fuels are developed substantially, in which case the gap in costs between the scenarios would increase.

Diminishing learning rates overtime were introduced to incorporate the fact that learning is tougher to obtain in the long run. The overall costs of the three scenarios as a share of the cumulated GDP over the period 2009–2050 are reproduced in Fig. 5.16b for rates at which learning diminishes ranging from 5 to 15 %. This figure indicates that the *business as usual* scenario is the least preferred scenario if a low rate at which learning diminishes applies. The other two scenarios rapidly become more expensive as this rate increases. Taken together, the sensitivity analysis indicate that the incremental cost of opting for the *renewable energy* scenario over the *business as usual* scenario could be between −0.4 and 1.5 % of the cumulated GDP. These results are in line with the findings of more sophisticated models (Edenhofer et al. 2006).

5.4.6 Concluding Remarks on Chapter 5

This last chapter has been an attempt to describe which energy technologies may be the most important ones in the global energy mixture of the future, more specifically the first half of the present century. At the same time we made an estimate of the relative cost of the TPES of a few carefully selected scenarios. It relied on a comparative static exercise of three scenarios representing different plausible futures. Future energy prices have been modeled using an innovative experience curve with diminishing learning rates. Our analysis shows that opting for the renewable energy scenario over the business as usual scenario would cost between −0.4 and 1.5 % of the cumulated GDP and that using nuclear energy can reduce this difference in costs.

Some aspects have been omitted here due to the absence of recognized methods to estimate their costs. The externality costs resulting from the use of various energy sources and the potential costs emanating from climate change (Edenhofer et al. 2011; Mideksa 2010; Stern 2007) and air pollution (Bollen et al. 2009) are not included in the *business as usual* scenario. Conversely, the cost of treating for intermittency and non-dispatchability issues, the cost of transforming electricity into a valuable fuel for the transportation sector, the cost of land area allocation and the cost associated to the construction of the infrastructure needed for storing energy and transporting electricity have not been included in the *nuclear* and the *renewable energy* scenarios. Although these categories of costs are likely to be compensated by similar existing costs in the fossil sector to some extent, these are important as they can impact the final cost of each scenario.

Nevertheless, in a future without significant discovery of fossil fuel reserves and no major technological breakthrough on the renewable energy side, the present analysis shows a remarkably small difference between the costs of the three energy scenarios. A transition to fully global renewable energy production is thus economically and technically possible. The cumulative capacities needed to achieve these price reductions do however rely on a large scale global development process which is difficult to foresee as long as energy policies are developed locally, since these tend to almost always favor the instant cheapest option.

5.5 Exercises

1. **Learning Rate and Cost Predictions**
   Consider the following data, compiling both the cumulative wind power capacity and the cost of new installations, 1990–2004.

   The second column indicates the cumulative installed wind power capacity in a country. The third column indicates the cost per kWh achieved by new installations in a given year.

	Cumulated capacity MW	Price Eurocents/kWh
1990	1,742	7.5
1991	1,983	7.46
1992	2,321	7.38
1993	2,801	7.3
1994	3,531	6.5
1995	4,821	6.0
1996	6,104	5.8
1997	7,636	5.6
1998	10,153	5.2
1999	13,494	4.8
2000	17,357	4.6
2001	24,444	4.4
2002	31,248	4.35
2003	39,431	4.3
2004	47,620	4.2

(a) Calculate the progress ratio. Explain what this progress ratio means for wind power.
(b) Calculate the average growth rate between 1990 and 2004.
(c) Make predictions on future costs. What are the predicted costs of wind power in Denmark for 2006 and 2010?
(d) In 2006, the real price per kWh is 5.2 Eurocents/kWh. Were your predictions accurate? If not, what can be a reason for such 'failure'? What does it tell you about the approach chosen?

2. **Externalities**
You are provided with the following information:

Label	Number	Unit
Plant economic life	40	years
Discount rate	10	%
Capital costs	409	Euro/kW
Fixed O&M costs	1.03	Euro/MWh
Fuel costs	15.68	Euro/MWh
Carbon cost	50	Euro/Tonne CO_2
Escalation rate	2.5	%
Carbon content	98.3	kgCO_2/GJ
Plant's mass-to-heat efficiency	46	%

In addition, you know that the plant is producing at full capacity during 7,446 hours per year.

(a) What is the LCOE for that plant?
(b) How does the LCOE change if you integrate the cost of emitting carbon dioxide?

3. **New Energy Mix-Europe**
   The average European total power consumption per citizen is about 5 kW, number of citizens are about and can be divided into three main consumption cathegories: 22 % transport (i), 35 % residential and commercial (ii) and 43 % industry (iii).
   (a) Suggest and argue for how much the contribution can be reduced in the transportation sector within a future more energy efficient society which sustains about the same standard of living as of today.
   (b) Suggest and argue for how much the contribution can be reduced in the residential and commercial sector within a future more energy efficient society which sustain about the same standard of living as of today.
   (c) Suggest and argue for how much the contribution can be reduced in the industry sector within a future more energy efficient society which sustain about the same standard of living as of today.
   (d) Calculate the total necessary power consumption per capita of Europe after this operation has been enforced.
   As a result of the present political and climate situation Europe decides to initiate an extremely aggressive non-fossil robust energy supply plan based on known non-fossil energy supply which should be operative and replace all fossil energy use in Europe within 20 years. It is assumed that the population remains stable at 711 million people and the average land area per person is about 14,000 m^2
   (e) Which energy sources and at which fraction of the total will you have in this energy mix? Justify how and why the chosen amount of each energy source is included.

References

Bollen, J., van der Zwaan, B., Brink, C.: Local air pollution and global climate change: A combined cost-benefit analysis. Resour. Energ. Econ. **31**(3), 161–181 (2009). doi:10.1016/j.reseneeco.2009.03.001

Edenhofer, O., Pichs-Madruga, R., Sokona, Y., Seyboth, K., Matschoss, P., Kadner, S., Zwickel, T., Eickemeier, P., Hansen, G., Schlomer, S., von Stechow, C.: Special report on renewable energy sources and climate change mitigation. Intergovernmental Panel on Climate Change, Working group III - Mitigation of Climate Change (2011)

Edenhofer, O., Lessmann, K., Kemfert, C., Grubb, M., Koehler, J.: Induced technological change: exploring its implications for the economics of atmospheric stabilization: synthesis report from the innovation modeling comparison project. Energ. J. **27**, 57–108 (2006). doi:10.5547/01956574.34.4.10

EPIA.: Market report 2011. European Photovoltaic Industry Association (2012)

EPI: World Average Photovoltaic Module Cost Per Watt, 1975–2006. Earth Policy Institute, Washington (2007)

EPIA/Greenpeace: Solar generation 6—solar photovoltaic electricity empowering the world. European Photovoltaic Industry Association, Brussels (2011)

Ferioli, F., Schoots, K., van der Zwaan, B.C.C.: Use and limitations of learning curves for energy technology policy: a component-learning hypothesis. Energ. Policy **37**(7), 2525–2535 (2009). doi:10.1016/j.enpol.2008.10.043

Helm, D.: The Carbon Crunch: How we're Getting Climate Change Wrong—and How to Fix it. Yale University Press, New Haven (2012)

IEA: Energy Balances of Non-oecd Countries 2011. OECD Publishing, Paris (2011a). doi:10.1787/energy_bal_non-oecd-2011-en

IEA: World Energy Outlook 1994. OECD Publishing, Paris (1994)

IEA: World Energy Outlook 2012. OECD Publishing, Paris (2012c)

IEA: World Energy Balances. OECD Publishing, Paris (2012a)

Jamasb, T.: Technical change theory and learning curves: patterns of progress in electricity generation technologies. Energ. J. **28**(3), 51–72 (2007)

Kahouli-Brahmi, S.: Technological learning in energy-environment-economy modelling: a survey. Energ. Policy **36**(1), 138–162 (2008). doi:10.1016/j.enpol.2007.09.001

Kolstad, C.D.: Environmental Economics. Oxford University Press Inc, New York (2000)

Krohn, S.: Wind Energy Policy in Denmark Status 2002. Danish Wind Industry Association, Denmark (2002)

Krohn, S., Morthorst, P.-E., Krohn, S., Morthorst, P.-E., Awerbuch, S.: The Economics of Wind Energy. European Wind Energy Association, Brussels (2009)

MacKay, D.J.C.: Sustainable Energy—Without the Hot Air. UIT, Cambridge (2009)

McDonald, A., Schrattenholzer, L.: Learning rates for energy technologies. Energ. Policy **29**(4), 255–261 (2001). doi:10.1016/S0301-4215(00)00122-1

Mideksa, T.: Economic and distributional impacts of climate change; the case of ethiopia. Global Environ. Chang. **20**(2), 278–286 (2010). doi:10.1016/j.gloenvcha.2009.11.007

Neij, L.: Cost development of future technologies for power generation—a study based on experience curves and complementary bottom-up assessments. Energ. Policy **36**(6), 2200–2211 (2008). doi:10.1016/j.enpol.2008.02.029

REN21: Renewables 2010 global status report. REN21 Secretariat (2010)

Roth, I.F., Ambs, L.L.: Incorporating externalities into a full cost approach to electric power generation life-cycle costing. Energy **29**(12–15), 2125–2144 (2004). doi:10.1016/j.energy.2004.03.016

Stern, N.: The Economics of Climate Change: The Stern Review. Cambridge University Press, Cambridge (2007)

Sullivan, J.L., Gaines, L.: A review of battery life-cycle analysis: state of knowledge and critical needs (2010)

Tamas, M.M., Shrestha, S.O.B., Zhou, H.: Feed-in tariff and tradable green certificate in oligopoly. Energ. Policy **38**(8), 4040–4047 (2010). doi:10.1016/j.enpol.2010.03.028

Timilsina, G.R., Kurdgelashvili, L., Narbel, P.A.: Solar energy: markets, economics and policies. Ren. Sust. Energ. Rev. **16**(1), 449–465 (2012). doi:10.1016/j.rser.2011.08.009

UN: World Population to 2300. United Nations Department of Economic and Social Affairs/Population Division, New York (2003)

United States Census Bureau. World population: 1950–2050, December 2008. URL http://www.census.gov/population/international/data/idb/worldpopgraph.php

Wright, T.P.: Factors affecting the cost of airplanes. J. Aeronaut. Sci. **3**(2), 122–128 (1936)

WWEA: World markets recovers and sets a new record: 42 GW of new capacity in 2011, total at 239 GW. World Wind Energy Association, Bonn (2012)

Author Biography

Patrick A. Narbel obtained his Ph.D. from the Norwegian School of Economics. He has researched the economics of energy, the efficiency of policy instruments facilitating the deployment of intermittent renewables and future prospects of these technologies. Patrick is a Senior Advisor at Sund Energy and a Council Member of the International Association for Energy Economics.

Jan Petter Hansen is a professor of the University of Bergen, Norway and adjoint professor of the the Norwegian School of Economics. At both places he is teaching basic physics related to energy. He has published more than 100 scientific papers and is currently a board member of he Norwegian Physical Society (NFS).

Jan R. Lien is professor emeritus from the University of Bergen, Norway. He has been teaching and doing research in both nuclear physics and petroleum technology and has several research terms abroad. He is a member of the Society of Petroleum Engineers (SPE) and the Norwegian Physical Society (NFS).

P. A. Narbel et al., *Energy Technologies and Economics*,
DOI: 10.1007/978-3-319-08225-7,

MIX
Papier aus verantwortungsvollen Quellen
Paper from responsible sources
FSC® C105338
FSC
www.fsc.org

If you have any concerns about our products,
you can contact us on
ProductSafety@springernature.com

In case Publisher is established outside the EU,
the EU authorized representative is:
Springer Nature Customer Service Center GmbH
Europaplatz 3, 69115 Heidelberg, Germany

Printed by Libri Plureos GmbH
in Hamburg, Germany